Fausto José Fernandes Dias
Flávia Gonçalves Fernandes

Arduino-Programmed Catapult for Oblique Launch Study

Fausto José Fernandes Dias
Flávia Gonçalves Fernandes

Arduino-Programmed Catapult for Oblique Launch Study

A Teaching Tool to Explore Math and Physics Concepts

ScienciaScripts

Imprint

Cover image: www.ingimage.com

This book is a translation from the original published under ISBN 978-620-6-76232-4.

Publisher:
Sciencia Scripts
is a trademark of
Dodo Books Indian Ocean Ltd. and OmniScriptum S.R.L publishing group

120 High Road, East Finchley, London, N2 9ED, United Kingdom
Str. Armeneasca 28/1, office 1, Chisinau MD-2012, Republic of Moldova, Europe
Managing Directors: Ieva Konstantinova, Victoria Ursu
info@omniscriptum.com

Printed at: see last page
ISBN: 978-620-8-52131-8

ACKNOWLEDGMENTS

To God, "For from him and through him and to him are all things. To him, therefore, be glory for ever. Amen!" (Romans 11.36)

To my parents Sadoc Ramiro Dias (*in memoriam*) and Margarida Terezinha Fernandes for the education and values they passed on to me and for everything they gave up in their lives so that I could exist.

To my wife Pollyana and children Davi and Elisa for the support they have given me throughout this journey and for the happy moments they have given me and for the many more come.

To my teacher and advisor Flávia, for the lessons given with impressive organization and clarity, and for the words of motivation and encouragement during the writing of this work.

To all those who conceived and put into practice this specialization course.

To that teacher whose name I can't remember and who perhaps never exchanged "good morning" and who, unpretentiously, shared the announcement of the registration for this specialization in a *WhatsApp* group and thus contributed to the existence of this work.

SUMMARY

This work aims to build an Arduino-controlled catapult that can launch objects at different angles and distances, as well as providing a practical environment studying oblique launching. A didactic sequence consisting of 18 hours of lessons is presented, with an innovative proposal for teaching mathematics and physics in high school: the use of programming and robotics concepts to study oblique launching. The step-by-step construction of a specific catapult model without programming or automation is presented. Basic concepts of electronics and programming are covered through interactive simulations of three platforms: *PheT Interactive Simulations*, *TinkeraCad* and the Arduino integrated development environment. It also describes a roadmap for programming and automation. The proposal aims to introduce the concept of oblique launching in a practical and fun way. During the course of the lessons, launches are made, data is collected such as launch angle and distance reached by the projectile, and the results obtained are analyzed. At the end of this work, it is hoped that teachers will use the proposed didactic sequence and understand that mathematics, physics and robotics complementary and interconnected.

Keywords: Mathematics. Programming. Robotics.

ABSTRACT

This work aims to build an Arduino-controlled catapult that can launch objects at different angles and distances, in addition to providing a practical environment for studying oblique throwing. A didactic sequence consisting of 18 hours/classes is presented with an innovative proposal for teaching mathematics and physics in high school: the use of programming and robotics concepts to study oblique throwing. A step-by-step guide to building a specific catapult model without programming or automation is presented. Basic electronics and programming concepts are covered through interactive simulations on three platforms: PheT Interactive Simulations, TinkeraCad and the Arduino integrated development environment. It also describes a roadmap on how to do its programming and automation. The proposal aims to introduce the concept of oblique launch in a practical and playful way. During classes, launches are made, data collection such as launch angle and distance reached by the projectile and analysis of the results obtained. At the end of this work, it is expected that teachers will use the proposed didactic sequence and understand that mathematics, physics and robotics are complementary and interconnected.

Keywords: Mathematics. Schedule. Robotics.

SUMMARY

1 INTRODUCTION

The act of teaching requires teachers to be researchers who are eager to discover new methodological strategies. Proposing innovative situations is a necessary challenge. In most situations, teachers deal with the same age group at the start of each new school year. If the educator is not aware of the changes in society, they may be led to believe that their previous successful plans can be reused without updating. (GARCIA and LABRE, 2021, p. 48) state that "These changes have required individuals and institutions to renew and adapt, which has also occurred in the school space and in teaching practices.". It is therefore important for teachers to keep up to date and understand how new student minds are being formed.

The Alpha generation, as those born after 2010 are called, is made up of children and teenagers who were born and raised with the internet and sophisticated electronic devices at their disposal, as (GARCIA and LABRE, 2021, p. 44) describe this generation. Many handle these devices more easily than their parents and teachers. Aligning students' intimacy with these tools with a lesson plan that makes them actively participate in the proposed activities is a strategy that can work very well if it is well planned, managed and led by the teacher as a mediator of learning.

This paper proposes a didactic sequence that uses Educational Robotics as a didactic resource to study and carry out practical experiments on the Knowledge Object: Oblique Launching. Oblique Launching is provided for in the Curriculum Document for Goiás - High School (DCGO-EM) to be developed in the second bimester of the first year of high school in the Mathematics curriculum component by describing the learning objective: Use information about the vertex of the parabola, determining its relationship with oblique launches to establish a trajectory and/or maximum range value.

The Common National Curriculum Base establishes as one of its general competencies be developed throughout secondary education both Mathematics and Physics:

> Understand, use and create digital information technologies and communication in a critical, meaningful, reflective and ethical way in the various social practices (including school practices) in order to communicate, access and disseminate information, produce knowledge,

> solve problems and exercise leadership and authorship in personal and collective life. (BRASIL, 2018, p. 7)

Every new didactic proposal developed by an educator should have, or at least should have as its main motivation, the improvement of their practice with the aim of improving student learning. And this is exactly what motivated the choice of this topic: to research an innovative strategy that is not only challenging personal and professional development but also, and above all, exciting and contributes significantly to students' understanding, assimilation and mastery of skills.

The proposed didactic sequence will be developed alongside the construction of a simple catapult which will be connected to and controlled by an Arduino platform provided by the teacher. The students will program the code using *Tinkercad*. They will also make the structure of the catapult by hand using reusable materials.

2 OBJECTIVES

This project aims to contribute to the development not only of students but also of educators. Some of the objectives are listed below.

2.1 General Objective

- To build an Arduino-controlled catapult that can launch objects at different angles and distances, as well as providing a practical environment for studying oblique launching;

2.2 Specific objectives

- Apply mathematical concepts in real-world contexts by building and programming robots;
- Apply the basic concepts of electronics in practical experiments;
- Learn basic programming concepts;
- Understand basic concepts of robotics and programming;
- Understand what Arduino is and its application;
- Build a catapult using knowledge of mathematics, physics, programming and electronics;
- Develop problem-solving skills;
- Develop concentration and observation;
- Develop teamwork, communication and collaboration skills by carrying out projects that combine robotics and mathematical concepts;
- Developing interest in mathematics through practical and fun activities with robotics;
- Developing a catapult controlled Arduino programming;
- Stimulate creativity by building projects and carrying out practical experiments;
- Study the influence of variables such as launch angle, object mass and applied force on the distance traveled by the launched object, promoting an understanding of the physical principles involved in movement;
- Familiarize yourself with the Arduino IDE (*Integrated Development Environment*);
- Encouraging interdisciplinarity, integrating objects of knowledge from mathematics, physics and science and technology through projects involving robotics;

- Integrating the mathematical concepts of trigonometry, kinematics and vectors with the readings and analysis of the data obtained in the experiments with the catapult, reinforcing the interdisciplinarity between mathematics and physics;
- Interest in the study of robotics;
- Organize ideas using a more sophisticated way of thinking;
- Promote the development of logical thinking and problem solving through the application of mathematical concepts in robot programming and control;
- Carry out experiments to collect data on the launches made by the catapult, making it possible to analyze graphs of position, velocity and acceleration a function of time;
- Relate the angle of the catapult shot in oblique launches to establish the maximum range of the projectile;
- Use concepts learned in other areas of knowledge and experience to develop a project.

3 BACKGROUND

The proposal to develop a didactic sequence for building an Arduino-controlled catapult capable of launching objects at different angles and distances arises from the need to provide students with a practical and interdisciplinary environment for studying oblique launching.

In an educational context that is increasingly focused on integrating technology and innovative teaching practices, the use of Arduino as a control platform offers a unique opportunity to combine physics, mathematics and programming concepts in a concrete and motivating project. Building and programming the catapult not only allows students to apply theoretical knowledge in a practical way, but also encourages the development of essential skills such as logical reasoning, problem-solving and teamwork.

In addition, the subject of oblique launching is fundamental to the study of classical mechanics, covering concepts such as trajectory, velocity, acceleration and maximum range. By exploring these concepts in practice, students can better understand the relationship between theory and practical application, strengthening their understanding and interest in the subject of physics.

The possibility of adjusting the parameters of the catapult, such as the launch angle and the force applied, allows students to explore different scenarios and analyze the effect of these variables on the range and trajectory of the objects launched. This provides a dynamic and personalized learning experience in which students can carry out experiments, collect data and perform quantitative analyses to validate their hypotheses and conclusions.

In addition, the integration of Arduino technology offers an interdisciplinary approach, connecting physics and math concepts with programming and engineering. Students have the opportunity to develop skills in microcontroller programming, learning to write and debug code to control the operation of the catapult according to the desired specifications. This broadens the students' range of skills, preparing them for technological challenges of the 21st century.

Thus, the creation of a didactic sequence for building an Arduino-controlled catapult represents an innovative and effective pedagogical strategy for teaching mathematics and physics, promoting student engagement, contextualizing content and developing essential skills for their academic and professional training.

Developing a project involving the construction of a catapult programmed in Arduino for the study of oblique launching in mathematics and physics is highly justifiable for the reasons described below:

3.1 Encouraging interdisciplinary learning

The integration of mathematics and physics through a practical and engaging project like this gives students the opportunity to apply theoretical concepts in a real context, promoting a deeper and more meaningful understanding of the content.

3.2 Stimulating creativity and innovation

Developing the catapult using Arduino programming challenges students to think creatively by exploring innovative technological solutions for building the device, thus stimulating critical thinking and problem-solving.

3.3 Development of practical skills

By actively participating in building, programming and carrying out experiments with the catapult, students develop practical skills in electronics, programming, mechanics and data analysis, preparing them for real-world challenges.

3.4 Encouraging teamwork

Collaborative projects like this encourage cooperation between students, promoting teamwork, effective communication and the division of tasks, essential skills academic and professional success.

3.5 Contextualization of content

The possibility of observing in practice the mathematical and physical principles related to oblique launching makes learning more concrete and interesting, as well as demonstrating the applicability of these concepts in the real world.

In this way, the development of this project is justified not only by enriching the educational process, but also by preparing students to face future challenges with solid technical and conceptual skills. It is also a suggestion for other fellow educators of an innovative proposal that can be developed and perfected in pedagogical practice.

4 THEORETICAL REVIEW

A curious and important fact to note is that more than two decades ago there were already papers published on the subject of teaching through educational robotics, with a significant increase in the production of scientific articles along the same lines in the last decade.

(CAMPOS, 2017) tried to explain why common sense has it that technology is everywhere but at school, and pointed out some of the challenges to having a productive and significant presence in the use of technological resources, especially robotics, for teaching purposes:

> The obstacles related to the implementation of robotics in the regular curriculum seem to us to be the nature of the time required for robotics activities, the cost of the necessary equipment and the theoretical and practical training of the teacher for the correct handling of the equipment, as well as the articulation of theory and practice in the use of this technological resource. (CAMPOS, 2017, p. 2112)

At least in mainstream schools, it is very clear that the curriculum provides for a huge number of skills to be developed by students during the school term. It is up to the teacher to diagnose the most relevant skills in order to give them greater attention. In this way, time is a real challenge when implementing robotics activities, given that the teacher needs to make an effort and demonstrate a rare talent so as not to jeopardize the curriculum proposed by the Common National Curriculum Base (BNCC) to the detriment of robotics. It is also noticeable that many public schools still have to worry about investing the money they receive in the infrastructure of their building and equipment and services such as computers and the internet, which makes it unfeasible in many cases to purchase more expensive equipment such as robotics.

An important consideration made by (CAMPOS, 2017, p. 2119) is that robotics itself is not what will provide an advance in the teaching-learning process, but rather a reformulation of the curriculum in order to integrate robotics with learning theories.

(CARVALHO and AGUIAR, 2018) developed an automated catapult with the Arduino Platform for the study of oblique launch in the teaching of physics at High School and concluded that "the proposed project was successful, obtaining effective results, related to oblique launch, the whole system is low cost because it has simple and accessible materials throughout its manufacture." They also reported how easy it was to carry out the calculations using the *Tracker* app.

However, (CARVALHO and AGUIAR, 2018) did not clearly describe the engagement obtained with the students in carrying out the project or the ease or difficulty faced in carrying out this activity. They also made an effort to emphasize what was achieved and built, to the detriment of how the stages of the project were carried out. , although the topic and even one of the products resulting from their article (the automated catapult with the Arduino Platform) are very similar to the one proposed in this research project, there are still aspects to be explored and described, such as, above all, the time the teacher will have to allocate to carrying out the project.

5 METHODOLOGY

Initially, specific bibliographic searches were carried out to verify the existence of articles or published works related to the construction of an automated or programmed catapult for teaching mathematics and/or physics. The only published work found was by (CARVALHO and AGUIAR, 2018), which presents the study of oblique launching using *Tracker* software and a catapult model programmed in Arduino, but without describing the construction process in detail.

A lot of research went into finding a catapult model that was easy to build and whose materials were recyclable and easy for teachers and students to buy. After the research, a prototype of the chosen model was built by the author to see how easy it was to make.

Using a descriptive and exploratory methodology, this work describes a teaching sequence of eighteen lessons for the third grade of secondary school. It details the steps and materials needed to build the catapult model chosen by the author without automation. It also describes the procedures and materials needed to automate the catapult and the use of platforms such as *Phet Interactive Simulations*, *TinkerCad* and the Arduino IDE as important tools for leveling up the basic concepts electronics and programming needed for automation and programming.

Within the didactic sequence presented, it is also proposed to carry out practical experiments of real launches with the catapult built and to compare these results with those found in the interactive oblique launch simulation that uses ideal conditions. Within each lesson, the students are asked questions stimulate logical-critical reasoning.

6 CURRICULUM STRUCTURE

6.1 Theme

Educational robotics for teaching mathematics.

6.2 Stage of education

Third grade.

6.3 Thematic unit

This didactic sequence addresses the following thematic units of the mathematics curriculum component: Numbers and algebra, and Geometry and measures.

6.4 General competences of the BNCC

"Throughout Basic Education, the essential learning defined in the BNCC should contribute to ensuring that students develop ten general competences, which embody, in the pedagogical sphere, learning and development rights." (BRASIL, 2018, p. 8).

See the definition of competence according to the BNCC:

> In the BNCC, competence is defined as the mobilization of knowledge (concepts and procedures), skills (practical, cognitive and socio-emotional), attitudes and values to solve the complex demands of everyday life, the full exercise of citizenship and the world of work (BRASIL, 2018, p. 8).

Among the ten general competencies defined by the BNCC, this work contributes to the development of at least five of them:

General Competence 01 - To value and use historically constructed knowledge about the physical, social, cultural and digital world in order to understand and explain reality, to continue learning and to contribute to building a fair, democratic and inclusive society.

General Competence 02 - Exercise intellectual curiosity and use the approach of the sciences, including research, reflection, critical analysis, imagination and creativity, to investigate causes, develop and test hypotheses, formulate and solve problems and create solutions (including technological ones) based on knowledge of the different areas.

General Competence 05 - Understand, use and create digital information and communication technologies in a critical, meaningful, reflective and ethical way in various social practices (including school practices) in order to communicate, access and disseminate information, produce knowledge, solve problems and exercise leadership and authorship in personal and collective life.

General Competence 09 - Exercising empathy, dialogue, conflict resolution and cooperation, showing respect for others and promoting respect for human rights, welcoming and valuing the diversity of individuals and social groups, their knowledge, identities, cultures and potential, without prejudice of any kind.

General Competence 10 - Acting personally and collectively with autonomy, responsibility, flexibility, resilience and determination, making decisions based on ethical, democratic, inclusive, sustainable and solidarity principles.

6.5 Objects of Knowledge **1.** oblique launch;

2. Graphs of functions;

3. Polynomial functions of the 2nd degree (quadratic function);

4. Critical points of a quadratic function: concavity, points of maximum or minimum.

6.6 BNCC skills

The BNCC establishes skills to be developed by students during each stage of education. Figure 1 below illustrates how to read the codes that identify the skills.

Figure 1 - Reading the BNCC skill codes

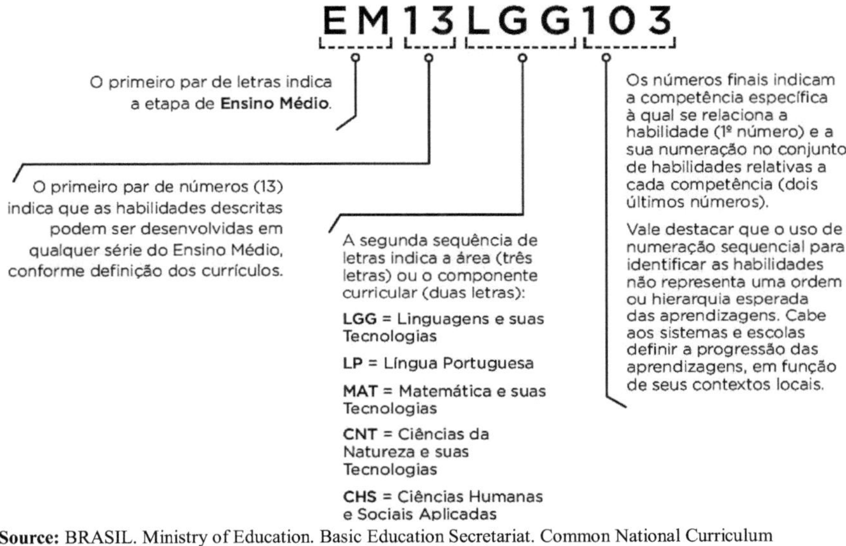

Source: BRASIL. Ministry of Education. Basic Education Secretariat. Common National Curriculum Base.

Education is the foundation. Brasília: MEC; SEB 2018. p. 34.

Listed below are some of the skills that can be developed in students who actively participate in the proposed didactic sequence.

•(EM13MAT315) Investigate and record, using a flowchart when possible, an algorithm that solves a problem;

•(EM13MAT405) Use the initial concepts of a programming language to implement algorithms written in ordinary and/or mathematical language;

•(EM13LGG702) Evaluate the impact of Digital Information and Communication Technologies (DICTs) on the formation of the subject and their social practices, in order to make critical use of this media in practices of selection, comprehension and production of discourses in a digital environment;

•(GO-701B) Exploring DICTs in an ethical, creative, responsible and appropriate way for language practices in different contexts, relating their constituent elements to their applicability in the social environment in order to expand the possibilities of using these digital tools;

•(GO-702A) Use different languages, media and digital tools in collective and collaborative production processes and authorial projects in digital environments, identifying implicit and ambivalent elements and their discursive effects in order to analyze the implications for the critical use of TDICs;

•(GO-702D) Organize study situations and use reading procedures and strategies appropriate to the objectives and nature of the knowledge in question, consulting and comparing different sources, tools and search sites to expand the perspectives of meaning construction.

6.7 Specific competences/skills of the Curriculum Document for Goiás - High School Stage (DC-GOEM)

In the state of Goiás, the BNCC skills are broken down in the DC-GOEM into other more specific and detailed skills. For example, the code GO-EMMAT315A is a breakdown of BNCC skill 315, which should be developed throughout secondary school () in the Mathematics curriculum component (MAT).

•(GO-EMMAT315A) Understand the concept of a flowchart as a graphic representation of the sequence of steps in a process, reading and identifying its basic symbols (start/end, arrow, connector, among others) and types (block diagram, simple process diagram, functional diagram, horizontal diagram, vertical diagram, among others) to map information presented in situations, as well as organizing the processes, reasoning and steps for solving the problem.

•(GOEMMAT315B) Organize, by means of flowcharts, the processes, reasoning and steps for solving the problem by selecting and characterizing the data and information presented to investigate the sequence of steps in a process, which involves numerical knowledge, algorithm, among others.

•(GO-EMMAT315C) Investigate and record, by means of a flowchart, algorithms that solve problems, analyzing their structure, the rules involved, the reasoning, logical procedures and/or operations used, among others, to build models and solve problems in various contexts.

•(GO-EMMAT405A) Understand the basic idea of algorithms as finite sequence of steps (instructions), registering mathematical representations (algebraic, geometric, statistical, computational, among others) referring to everyday situations (routine or not) to organize the process and use initial programming language concepts in the implementation of algorithms written in ordinary and/or mathematical language.

•(GOEMMAT405B) Use initial programming language concepts in the implementation of algorithms written in ordinary and/or mathematical language, analyzing the results and their implications for solving problems involving algebraic expressions, functions and/or mathematical algorithms, among others.

•(GO-EMMAT405C) Solve problems involving algebraic expressions, functions and mathematical algorithms, using initial programming language concepts to find and propose solutions in a variety of social contexts.

6.8 Learning objectives

•Apply mathematical concepts in real-world contexts by building and programming robots;

•Apply the basic concepts of electronics in practical experiments;

•Learn basic programming concepts;

•Understand basic concepts of robotics and programming;

•Understand what Arduino is and its application;

•Build a catapult using knowledge of mathematics, physics, programming and electronics;

•Develop problem-solving skills;

•Develop concentration and observation;

•Develop teamwork, communication and collaboration skills by carrying out projects that combine robotics and mathematical concepts;

•Developing interest in mathematics through practical and fun activities with robotics;

•Developing a catapult controlled Arduino programming;

•Stimulate creativity by building projects and carrying out practical experiments;

•Study the influence of variables such as launch angle, object mass and applied force on the distance traveled by the launched object, promoting an understanding of the physical principles involved in movement;

•Familiarize yourself with the Arduino IDE (*Integrated Development Environment*);

•Encouraging interdisciplinarity, integrating objects of knowledge from mathematics, physics and science and technology through projects involving robotics;

•Integrating the mathematical concepts of trigonometry, kinematics and vectors with the readings and analysis of the data obtained in the experiments with the catapult, reinforcing the interdisciplinarity between mathematics and physics;

• Interest in the study of robotics;

• Organize ideas using a more sophisticated way of thinking;

• Promote the development of logical thinking and problem solving through the application of mathematical concepts in robot programming and control;

• Carry out experiments to collect data on the launches made by the catapult, making it possible to analyze graphs of position, velocity and acceleration a function of time;

• Relate the angle of the catapult shot in oblique launches to establish the maximum range of the projectile;

• Use concepts learned in other areas of knowledge and experience to develop a project;

6.9 Duration of activities

The activities described in this didactic sequence will last 18 hours.

6.10 Prior knowledge worked on by the teacher with the student

The didactic sequence proposed in this work is a suggested introduction to the study oblique launching. Although there is no need for students to already know about this topic, it is important that students are able : read, write and interpret information and written commands.

Basic knowledge of Arduino programming would make it much easier to understand the codes presented during the didactic sequence, but is not necessarily essential for their development and use.

7 DESCRIPTION OF THE DIDACTIC SEQUENCE

Before starting to describe the didactic sequence lesson by lesson, the following are the general guidelines for the teacher to make one of the main products of this didactic sequence: the catapult.

7.1 General guidelines

The materials needed to build the catapult without the need for automation/programming are described below, followed by a separate list of the materials used to program and automate the catapult.

The materials used to program the Arduino and automate the catapult can be purchased by the teacher in at least three different ways: out of their own pocket, through a donation or by the school itself. In order for the purchase to be made by the school itself, the teacher must present this action to the school management in good time so that it can be registered in the action plan of one of the financial resources that the school receives.

7.1.1 Materials needed to build the catapult without automation/programming

The following materials were used to make the catapult shown:

Figure 2 - Twelve popsicle sticks measuring approximately 12 cm

Source: The author

Figure 3 - Hot glue gun and its glue stick

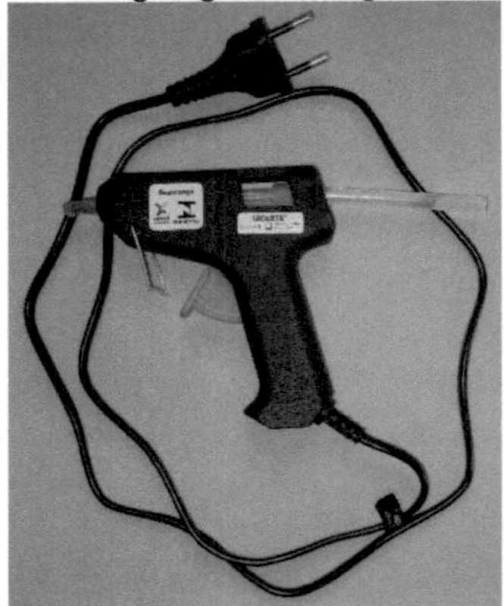

Source: The author

Figure 4 - A barbecue skewer

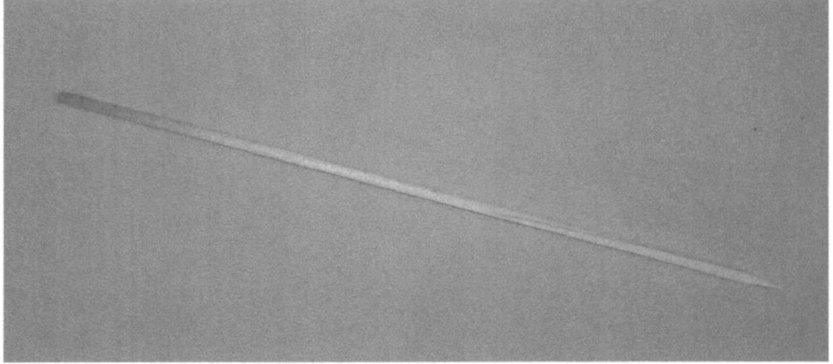

Source: The author

Figure 5 - Two 6cm x 5cm cardboard rectangles

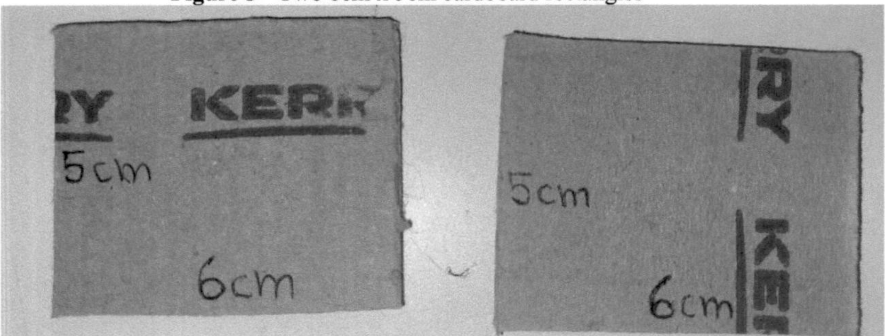

Source: The author

Figure 6 - A rubber band (a garter used to tie pamonhas for the Goianos)

Source: The author

Figure 7 - Two PET (polyethylene terephthalate) bottle caps

Source: The author

Figure 8 - Line, a stylus and a ruler

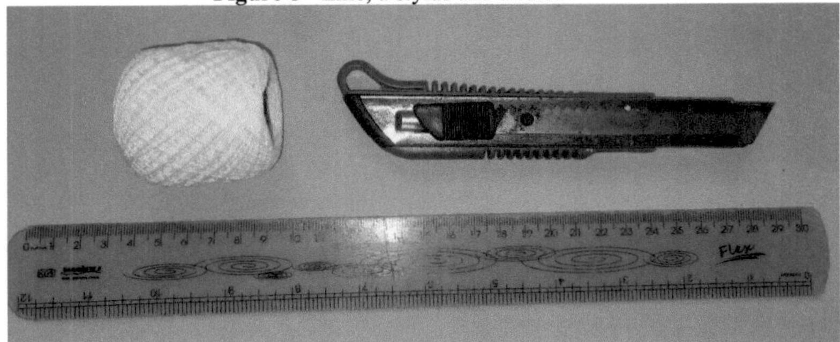

Source: The author

Figure 9 - A protractor from 0 to 180 degrees

Source: The author

Figure 10 - No. 120 sandpaper to finish off the cut pieces of toothpick

Source: The author

7.1.2 Step-by-step construction of the catapult without automation/programming

It is very important that the teacher builds at least one catapult model to present to the students in the first lesson of this sequence and requests the necessary materials already in the first or second lesson to encourage the students to build the catapult in working groups four students in the following lessons.

There are countless catapult models on the internet. This model was chosen and adapted for this sequence because it uses materials that are easy to buy, both for the school and for the teacher and students.

Figure 11 - Catapult ready without automation

Source: The author

The step-by-step construction of the chosen catapult model is detailed below.

7.1.2.1 Step 1: Cut out the popsicle sticks and barbecue skewers

Using a ruler, pen and stylus, the measurements shown in the figure below are marked out and the sticks cut out. Five sticks will be used whole, without the need for cutting. The whole sticks in the picture below are about 12 cm long and 0.8 cm wide. There may be variations in size from one stick to another and it is important to note this so as not to use sticks of different sizes and compromise the structure. The picture below also shows a piece of barbecue skewer cut to a length of 5 cm.

This is when the 120 grit sandpaper comes into play to give a better finish to the cuts. Take care to hold the sticks of the same size together when sanding to avoid sanding them to different sizes.

Figure 12 - Popsicle cut with measurements in centimeters

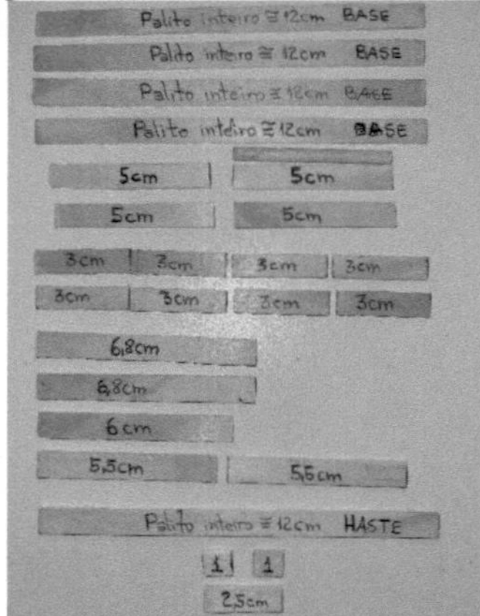

Source: The author

7.1.2.2 Step 2: Assemble the first stage of the base

Two 5 cm pieces of toothpick are glued on top of each other with hot glue and then glued to the ends of two of the base's whole toothpicks, forming a 90° angle as shown in the figures below.

Figure 13 - First stage of the assembled base

Source: The author

Figure 14 - Another view of the first stage of the assembled base

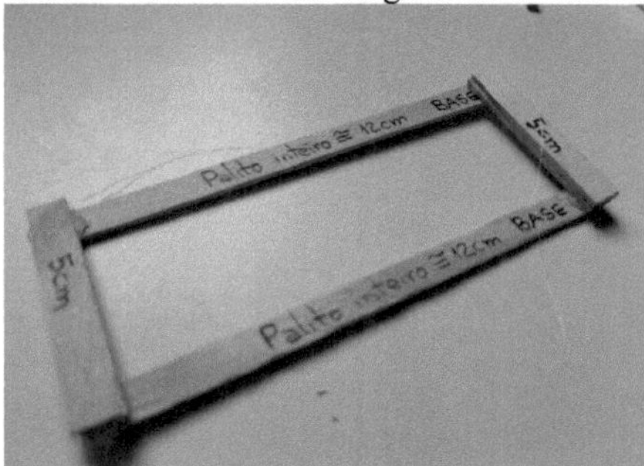

Source: The author

7.1.2.3 Step 3: Finish the base of the catapult

The pieces of toothpick cut to a length of 3 cm are also glued together two by two with hot glue as shown in the picture below.

Figure 15 - 3 cm pieces glued together two by two

Source: The author

Mark off half of the base sticks that have already been glued together and leave a gap of 0.5 cm. It is in this space that the catapult's shaft will be fitted and should rotate freely. The barbecue skewer used as the shaft is 0.4 cm in diameter, so the 0.5 cm space allows it to rotate freely. It is important that the teacher takes care to check that this suggested measurement also fits their project according to the materials they and their students have collected. A safe way of ensuring this is to use the material that will be used for the catapult shaft as a measure. The marking described should look like the figure below.

Figure 16 - Detail of the spacing at the base where the shaft should rotate freely

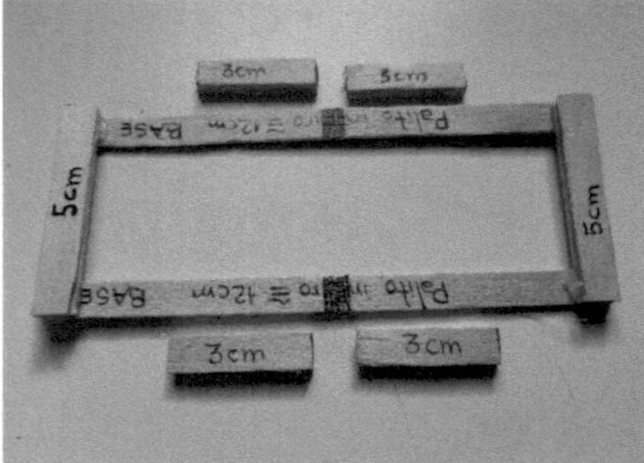

Source: The author

After this, the 3 cm pieces are glued together respecting the spacing left for the axle and taking care that the hot glue is not used excessively and takes up part of this space. If this happens, use the stylus to remove the glue that has dripped onto this part. See how it should in the picture below.

Figure 17 - 3 cm pieces glued to the base

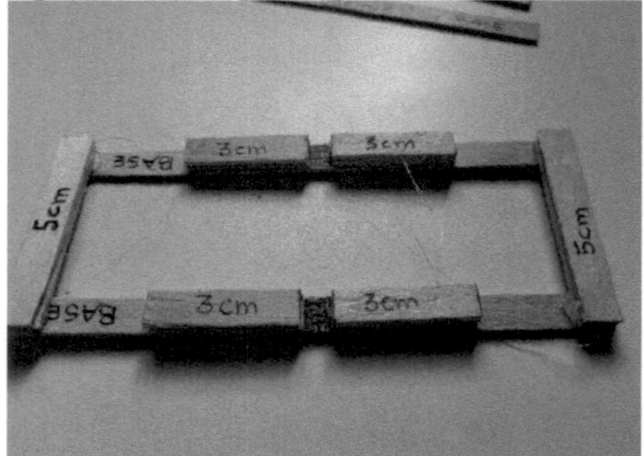

Source: The author

To finish off this step, the last two whole sticks of the base are glued on top of this structure, as shown in the image below. Once again, care must be taken to avoid hot glue occupying the space where the catapult shaft will be placed.

Figure 18 - Finished catapult base

Source: The author

Figure 19 - Catapult shaft positioned to check that it rotates freely

Source: The author

7.1.2.4 Step 4: Position the rod on the catapult shaft

After checking that the shaft of the catapult rotates freely in its position, use the stylus to make a hole the thickness of a toothpick in the shaft, as shown in the figure below.

Figure 20 - Detail of the hole in the catapult shaft

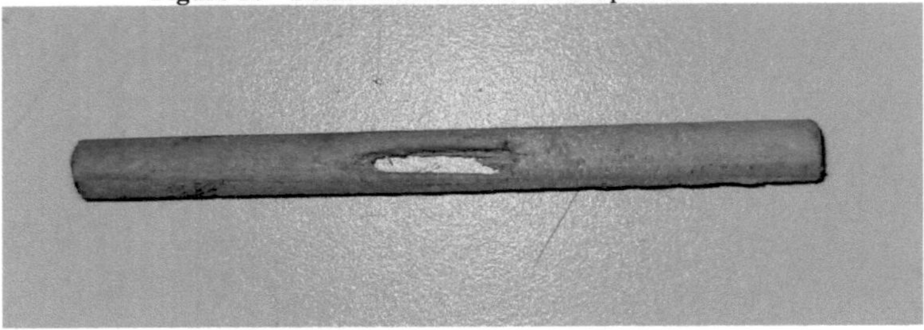

Source: The author

Then, using sandpaper no. 120 once again, you need to sand one end of the entire stick that will be the catapult's shaft so that it fits correctly into this hole in the axle. This hole is designed to improve the shaft's hold on the axle, but if it's difficult to drill this hole, the shaft can be glued directly to the axle without need for a hole. At this point it is not recommended to glue the rod to the shaft, just leave it in place. The figure below shows the rod fitted to the catapult shaft.

Figure 21 - Rod fitted to the catapult shaft

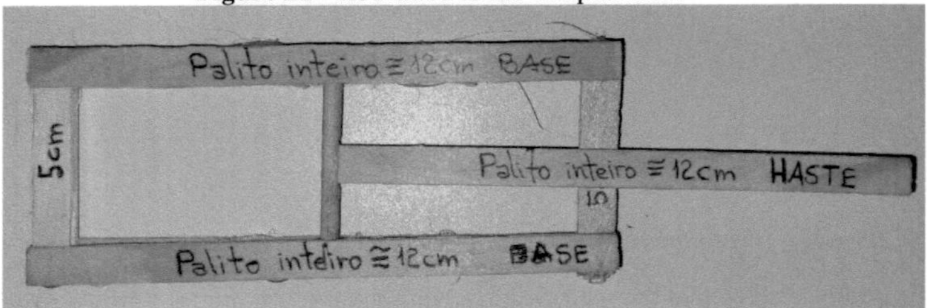

Source: The author

7.1.2.5 Step 5: Fitting the catapult launch limiter

Now it's time to use those pieces of toothpick cut to lengths of 5.5 cm, 5.5 cm and 6 cm to form the structure that will prevent the catapult shaft from exceeding the launch angle of 90º, the maximum launch angle. With the catapult shaft in place, glue the 5.5 cm pieces together, one on each side of the shaft and perpendicular to the base of the catapult. Then glue the 6 cm piece so that there is a small piece left on each side. This remnant will be used to secure the elastic that will be used. When you have finished this step, your catapult should look like the one in the image below.

Figure 22 - Mounted launch

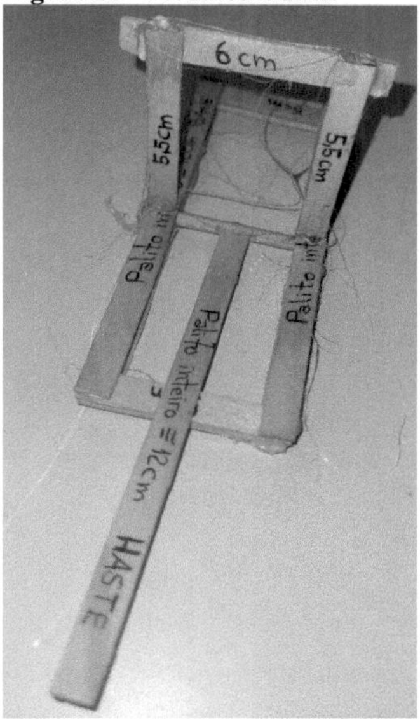

Source: The author

7.1.2.6 Step 6: Finalize the launch limiter

In order to give greater strength to this structure called the launch limiter in this article, it is necessary to use the only rigid polygon in mathematics: the triangle. To do this, use another 5 cm piece of toothpick to glue to the end of the base and then use the two 6.8 cm pieces to fit between the limiter and the 5 cm piece. Once they're in place, use plenty of hot glue to secure and strengthen the structure even more. At the end of this step, your catapult should look like the one shown in the picture below.

Figure 23 - Reinforced launch

Source: The author

7.1.2.7 Step 7: Finishing the catapult shaft

The 1 cm and 2.5 cm long pieces of toothpick will be used to lock the elastic and prevent it from sliding down the shaft when the catapult is armed. With the rod and shaft in place, lift the rod up to the launch stop and make a mark at the height of the stop of 0.5 cm where the rubber band will later be passed through, as shown in the image below.

Figure 24 - Marking on the rod to pass the elastic through

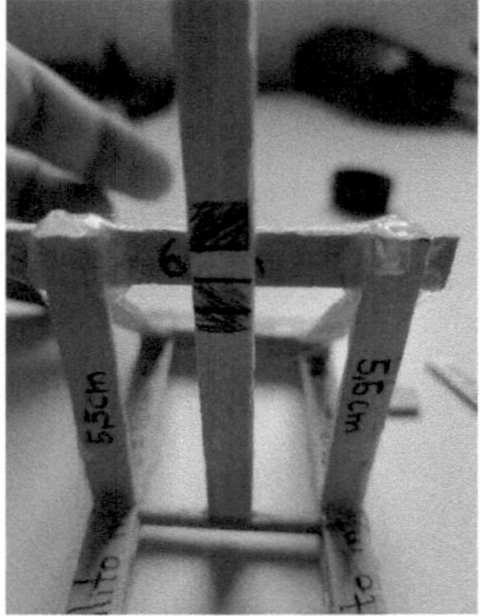

Source: The author

Once the markings have been made, it's time to glue the two 1cm pieces together as shown in the image below

Figure 25 - 1cm pieces glued to the stem

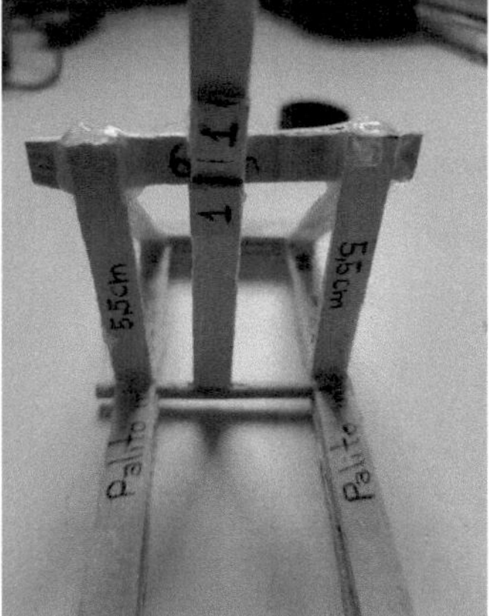

Source: The author

To finish this step, the 2.5 cm piece on top of the 1 cm pieces. The elastic should go through this hole once the hot glue has dried completely. See how it should in the picture below.

Figure 26 - 2.5cm piece glued over the 1cm pieces

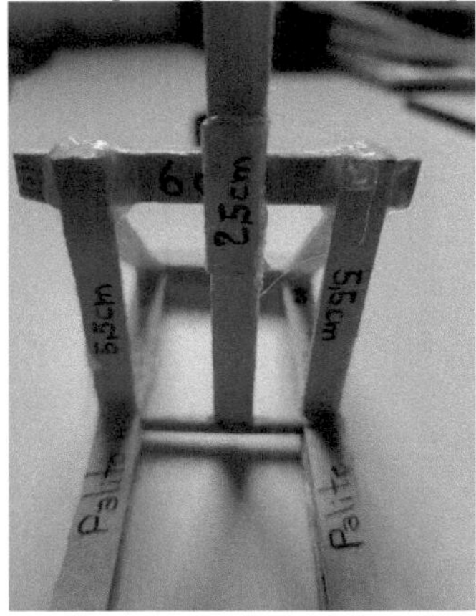

Source: The author

Now it's time to use one of the PET bottle caps, which should be glued to the end of the catapult shaft. You can also use hot glue to fit the shaft of the catapult and wait for the glue to dry before attaching the elastic.

Figure 27 - Soda cap glued to the stem

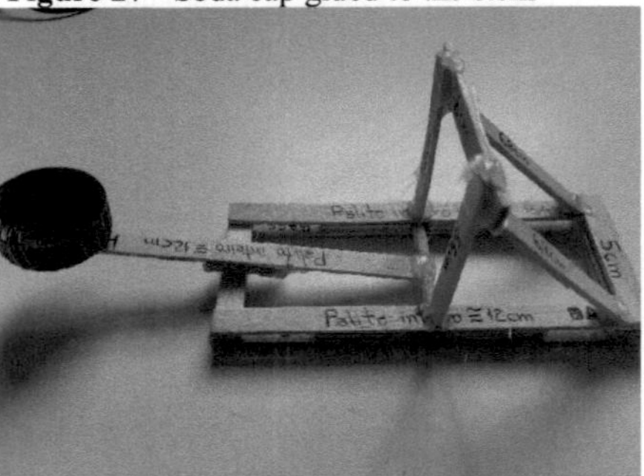

Source: The author

7.1.2.8 Step 8: Thread the elastic onto the rod and fit the release limiter

Now it's time to pass the elastic between those 1cm pieces that have been glued to the stem. To do this, you can use a barbecue skewer to help push the elastic through the gap. Be careful not to cut the elastic in this process. In this project, after passing the elastic through the hole, each end of the elastic went over the rod in the opposite direction and was fitted to one end of the 6cm piece used for the launch limiter. Take a look at the picture below. It's important to note that this finish will depend on the size of the elastic used, which can vary. If the elastic is still loose on the structure, you can go around the rod a bit more ensure a firmer hold.

Figure 28 - Detail of the rubber band in its final position

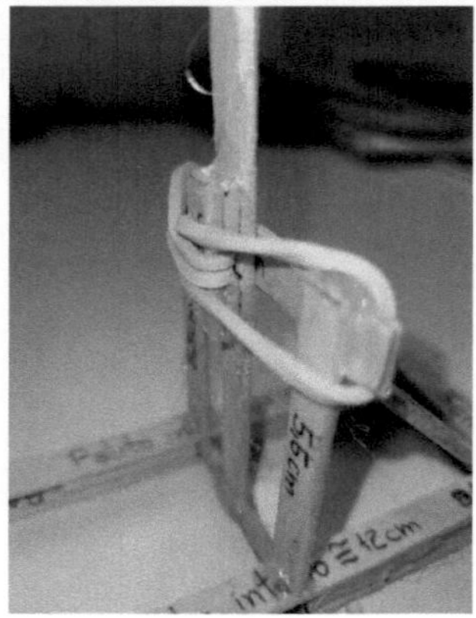

Source: The author

Figure 29 - Perspective view of the rubber band in its final position

Source: The author

At this stage, the catapult is ready to carry out launch tests and check its resistance and range with a 90° launch angle. I'm sure you and the students have already done these tests to see if it really works. Have fun!

7.1.2.9 Step 9: Glue the cardboard rectangle and mark the notable angles on it

Now it's time to use the two cardboard rectangles which will only work to limit the launch to smaller angles: 30°, 45° and 60°. Before marking the angle, you need to glue the cardboard to the catapult as shown in the figure below.

Figure 30 - Cardboard glued to one side of the catapult

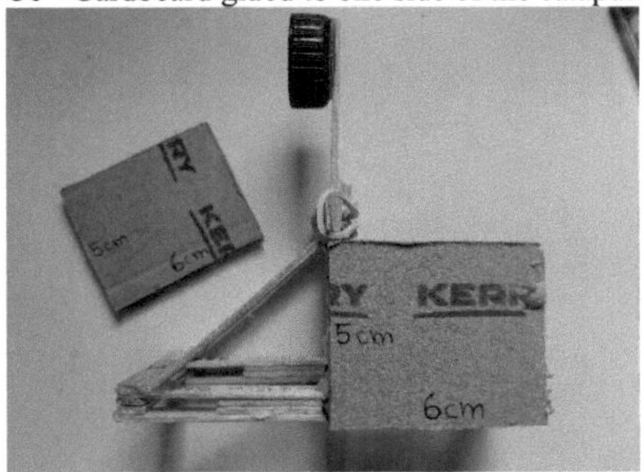

Source: The author

At this stage, be careful not to leave a piece of cardboard sticking out below the base of the catapult. If do, use a stylus to remove the excess.

Figure 31 - Cardboard glued to both sides of the catapult

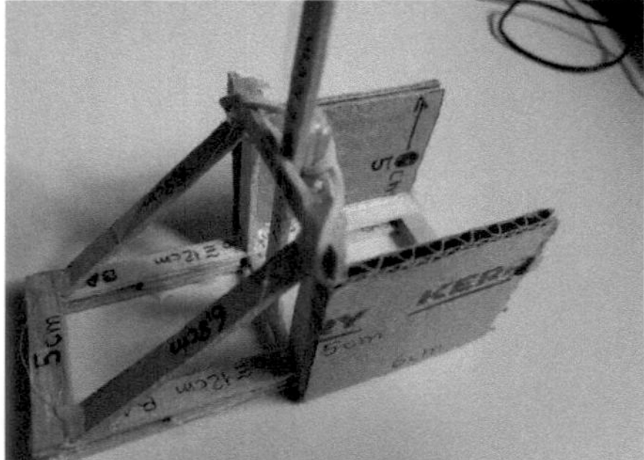

Source: The author

With the cardboard glued down, it's time to mark the angles we want to use in the tests during the lessons. At this point you can mark other angles besides the notable ones if you wish. An important detail is that the origin of the angles must be the axis of the catapult and not the bottom corner of the cardboard. You'll notice that the two hardly ever coincide. Because of the measurements used, it is recommended to make the markings on the inside of the protractor to ensure that the distance from the origin to the marking is the same on both sides of the cardboard. See how these angles are marked in the images below.

Figure 32 - Marking the notable angles on side 1 of the catapult

Source: The author

Figure 33 - Marking the notable angles on side 2 of the catapult

Source: The author

With the markings made, you can now use the barbecue skewer once again to make a hole at each marked angle and on side of the catapult.

Figure 34 - Piercing the cardboard with a barbecue skewer

Source: The author

The figure below shows the catapult with all the holes drilled in the notable angles and the remaining piece of the skewer that was used to drill the holes will still be used as a mobile limiter for launch angles less than 90°. If you want to launch at an angle of 30°, simply place the barbecue skewer in this position after setting up the catapult.

Figure 35 - Catapult without automation finished and ready for use

Source: The author

If you, the teacher, are afraid of Arduino programming or your school hasn't bought a robotics kit yet, or you don't even want to risk buying a robotics kit, you should know that this project can already be used normally in your practical lessons with students. However, if you're a teacher who's interested in going one step further and bringing something even more exciting to your students, follow the step-by-step instructions below for programming part where youll make this catapult arm and disarm itself using Arduino code.

7.1.3 Materials needed for Arduino automation/programming

The following materials and electronic components were used to make the catapult shown:

Figure 36 - An Arduino UNO or compatible board (Average cost: R$ 50.00)

Source: The author

Figure 37 - A USB cable for the Arduino board (Average price: usually comes with the Arduino board)

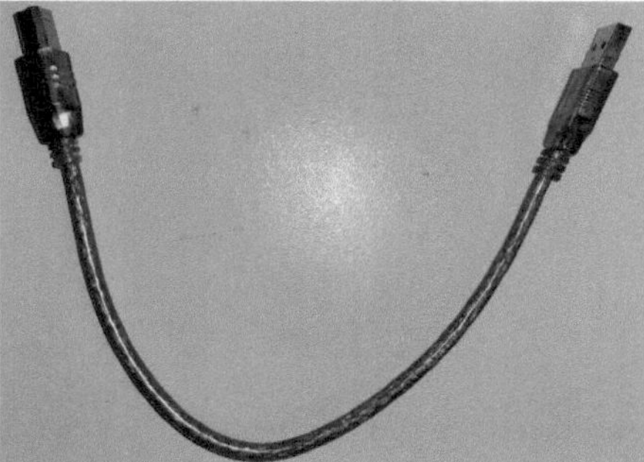

Source: The author

Figure 38 - A 9g SG 90 micro servo motor (Average cost: R$12.00)

Source: The author

Figure 39 - A 28BYJ-48 stepper motor + a UNL2003 stepper motor driver (Average cost: R$ 30,00)

Source: The author

Figure 40 - Thirteen 12 cm male jumper cables (Average cost: R$ 5.00)

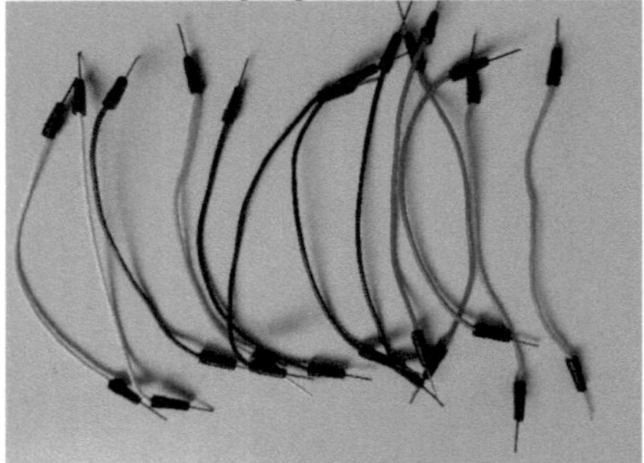

Source: The author

Figure 41 - Six 20 cm female jumper cables (Average cost: R$ 5.00)

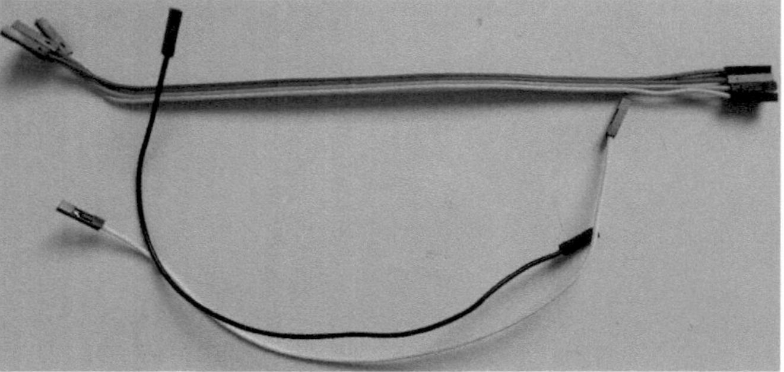

Source: The author

Figure 42 - A *protoboard* (Average cost: R$ 10,00)

Source: The author

Figure 43 - Two *push buttons* with cover (Average cost: R$ 2,00)

Source: The author

Figure 44 - **(**Optional) Two 5.5 cm rigid *jumpers* and two 2 cm rigid jumpers (Average cost: R$ 0.60)

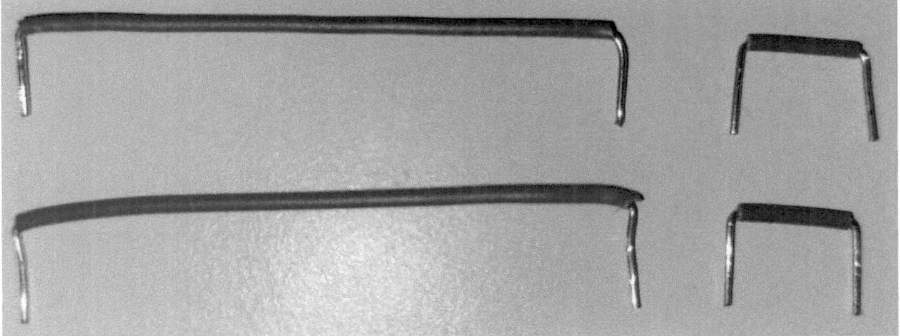

Source: The author

Figure 45 - (Optional) a pair of tweezers

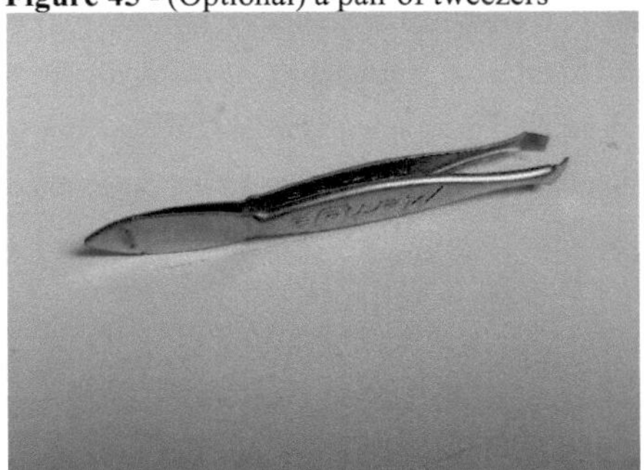

Source: The author

Figure 46 - 30 cm of string

Source: The author

Figure 47 - 30 cm of wood (*Pallet*)

Source: The author

Considering that the string and tweezers don't necessarily have to be expensive, as they are easy to find in almost every home, we estimate a total average cost of approximately R$114.60 (one hundred and fourteen reais and sixty cents) to purchase all the components needed to program and automate the catapult. It was also assumed that the computer/notebook used for programming would belong to the teacher or the school. It is worth remembering that it is possible to buy all the electronic components listed in kits, which would reduce the cost.

7.1.4 Step-by-step programming and automation of the catapult

Below are the steps required to carry out the programming and its integration with the catapult built in order finally obtain the automation of the catapult.

7.1.4.1 Step 1: Acquire basic knowledge of electronics and the components to be used

To carry out this step-by-step, it is important that the teacher has a basic knowledge of electronics and the use of the components listed above. However, if you don't have any knowledge, you can find many video tutorials on the internet that share knowledge and help you get started with this type of material. In the same way, you'll find lots of material and videos on Arduino programming.

7.1.4.2 Step 2: Download the Arduino IDE software on a computer/notebook

Once you know how each component works and can be programmed, you can make various adaptations to improve the code provided in APPENDIX A - ARDUINO CODE FOR PROGRAMMING AND AUTOMATING THE CATAPULT, which can be copied and pasted directly into the Arduino IDE software (Arduino Integrated Development Environment). Arduino IDE is free software used to program Arduino and other compatible boards. It offers a set of essential features for writing, analyzing and loading code into the board's microcontroller. This software can be downloaded free of charge from: <https://www.arduino.cc/en/software> Accessed on: July 1, 2024.

7.1.4.3 Step 3: Assemble the circuit.

In APPENDIX B - SCHEMATICAL VIEW OF THE CIRCUIT FOR CATAPULT AUTOMATION details how the stepper motor, servo motor and buttons must be connected to the protoboard and Arduino board for the code in APPENDIX A to work correctly.

The tweezers suggested among the materials help a lot when handling the cables, which are many and thin. When assembling the circuit, make sure that the jumper cables are properly connected. Sometimes the code doesn't seem to work properly because one or more cables get disconnected without you noticing. It is also worth pointing out that in order to run the code in the Arduino IDE software, the Arduino UNO board must be connected to the computer/notebook on which the software is running. It's very common, due to distraction, to forget to make this connection and imagine that there's a error in the code or that the cables are not in contact or are even damaged.

The function of the servo motor is to control the firing of the catapult, while the function of the stepper motor is to drive the catapult's throwing arm. These functions are controlled by their respective buttons (servo motor = button connected to port 5; stepper motor connected to port 6).

7.1.4.4 Step 4: Integration between the catapult and its automation

To complete the assembly of the catapult and finalize its automation, glue the catapult to the piece of wood used for the base as shown in the image below.

Figure 48 - Catapult glued to wooden base

Source: The author

Next, drill a very small hole in the PET bottle cap and toothpick of the throwing arm. This hole is for the string that will be tied to the stepper motor and will drive the catapult's throwing arm.

Now you need to position the servo motor so that it locks the catapult's throwing arm. Mark the correct position before applying the hot glue. Don't worry at this point about which way the servo motor will turn, as you can set this in the programming code. Take a look at the figure below which illustrates this step.

Figure 49 - Positioning the servo motor before gluing and detail of the hole in the cover

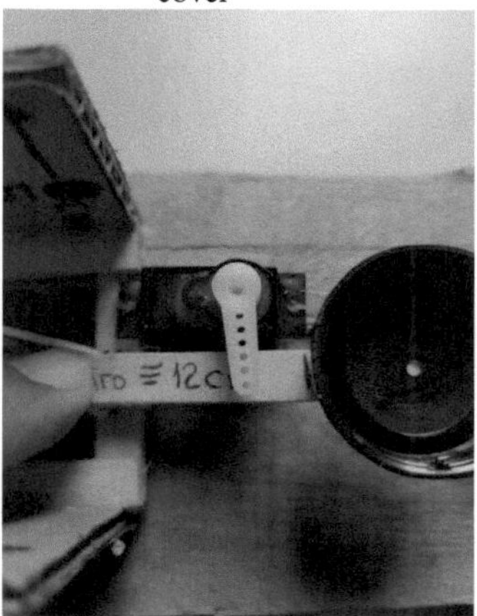

Source: The author

The second PET bottle cap is used to reduce the time taken by the stepper motor to drive the catapult. The first attempt was to tie the string directly to the stepper motor shaft. However, due to the length of the circumference the stepper motor shaft and its speed, it was necessary to find an alternative so that traction could be achieved in a shorter time. Therefore, the function of the cap glued to the stepper motor shaft is simply to increase the length of the circumference that will wrap the string. The following pictures show how this step turned out.

Figure 50 - Cap glued to stepper motor shaft

Source: The author

Figure 51 - Cap glued to the stepper motor shaft

Source: The author

The following figures show the servo motor and stepper motor positioned on the base of the catapult before being hot glued into a fixed position.

Figure 52 - Servo motor and stepper motor positioned

Source: The author

Once positioned and moving the catapult's throwing arm to ensure the ideal position, mark this position with a pen and apply plenty of hot glue as they should be well fixed due to the strain they will be subjected to by the elastic force exerted by rubber band.

Figure 53 - Stepper motor glued to the base

Source: The author

Once all the elements are in place, cut a piece of toothpick or skewer to act as a lock on the cap of the catapult's throwing arm. Tie one end of the 30 cm string to this piece of toothpick and pass it through the hole in the cap. The other end of the string should be tightly tied to the cap that has been glued to the stepper motor. See how it should look in the following pictures.

Figure 54 - String that will be connected to catapult and stepper motor

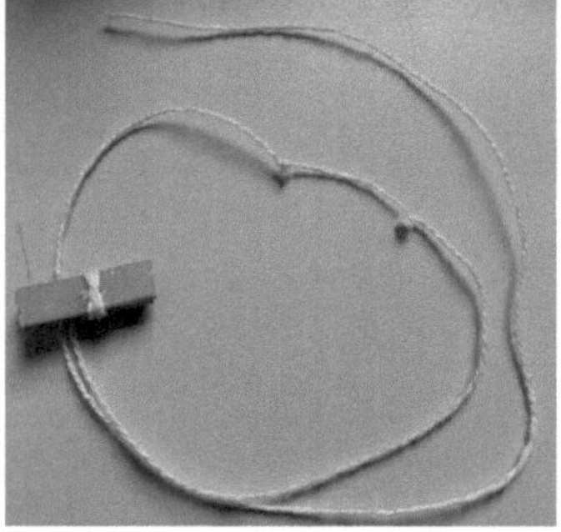

Source: The author

Figure 55 - Catapult ready to be connected to the *protoboard*

Source: The author

Figure 56 - Automated catapult with *jumpers* connected

Source: The author

Figure 57 - Automated catapult with *jumper* cables connected - side view

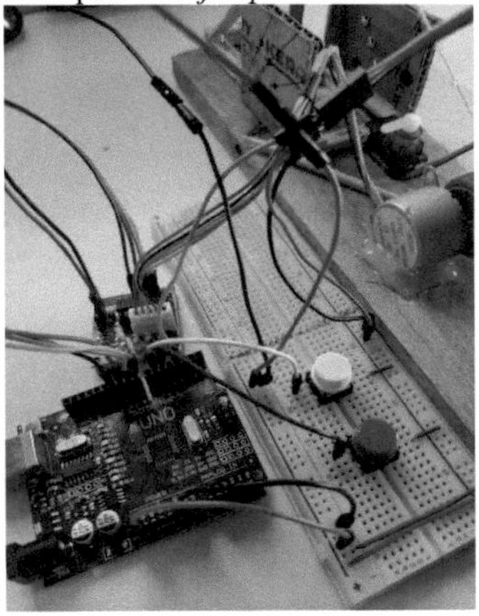

Source: The author

7.2 Classes 01 and 02 - Presentation of the catapult and Arduino's electronic components (1h 40min)

In the first lesson of this didactic sequence, it is essential that the automated catapult has already been developed by you, the teacher. This is the time to encourage student participation and find out what they already know about the topics to be covered. It is essential that the teacher captivates and arouses the students' interest in carrying out the activities that will be proposed during this sequence. To do this, it is important that students have contact with the product of the final objective: to build a catapult using knowledge of mathematics, physics, programming and electronics. It is therefore advisable for the teacher to take the finished Catapult project into the classroom and show the students how it works, as well as describing the stages in a detailed and pedagogically prepared step-by-step guide.

7.2.1 Specific objectives:

- To arouse interest in the study of robotics;
- Understand basic concepts of robotics and programming;
- Encourage a interdisciplinarity, integrating objects of knowledge of mathematics, physics and science and technology through projects involving robotics;

•Developing interest in mathematics through practical and fun activities with robotics;

7.2.2 Teaching resources:

•Notebook;

•Catapult programmed in Arduino;

•Chart;

•Chalk;

•Media projector or television set;

•Printed list of materials needed to build the catapult without automation.

7.2.3 Introduction - 25 min

When you enter the classroom with the catapult, all attention will be focused on it. Take the opportunity to ask questions and listen to the students. Some suggested questions:

Does anyone know what this is that I've brought today? What is it for? Has anyone seen it in a movie or history book? Depending on the class, open-ended questions that aren't directed at a specific person can go unanswered. This is where the teacher's role comes into play, by selecting those students who are most likely to answer and asking them targeted questions. You'll find that others will be motivated to answer and participate after the first.

7.2.4 Development - 25 min

If any student recognizes that the object shown is a catapult, take advantage of the answer to ask other related questions such as: What does a catapult do? It's expected that high school students will already know about the catapult and what it can do from movies and series they've watched or even digital games. Take references from films and series in which the catapult is used as a strategy to get closer to the students and connect with their reality. Give them space to cite their references too and show interest and curiosity in their answers.

7.2.5 Practical application - 30 min

Take the opportunity to briefly discuss the history of the catapult with students using videos and other resources. In the topic Suggested support materials, a video of almost 7 minutes in length from the Medieval Channel on YouTube is provided, which tells a little about the history and construction methods of one of the various types of catapult. In addition to the video, an article from Editora Abril's digital magazine Super interessante, which tells the story of catapults, is also suggested.

It is possible that the students will question and ask about the wires and components incorporated into the catapult. If they do, explain that the idea is to build a catapult with the students that will be launched automatically through programming. Make a few launches to demonstrate how it works, and let the students use the trigger and firing buttons.

7.2.6 Conclusion - 20 min

Introduce the specific objectives of these lessons and ask them to divide into groups four to find the materials they need to build the catapult without automation. You could also bring a printed list of these materials or write them on the board and ask the students to copy them down. If you think it necessary, to increase the students' commitment and engagement, suggest that at the end there be a competition in which the winning group will be the one that develops a catapult with the longest projectile launch range. Give the students a week to get the materials they need and end the lesson with the confidence that the work will be a success.

7.2.7 Suggested support materials

CANAL MEDIEVAL (Brazil). Catapults: Trebuchet or Trabuco - Medieval Siege Weapons Series. **YouTube**. Available at: https://www.youtube.com/watch?v=ZBQntekpfpI. Accessed on: July 2, 2024.

ONÇA, Fabiano. Catapult: the story of the invention that changed the history of warfare. **Super interesting**. São Paulo, October 30, 2019. Available at: https://super.abril.com.br/historia/a- mae-de-todas-as-guerras. Accessed on: July 2, 2024.

7.3 Lessons 03 and 04 - Algorithm for building the catapult and execution of steps 1, 2 and 3 (1h 40min)

The development of this lesson will depend on the students' response to acquiring the necessary materials. It is in this lesson that the project timetable will be developed and the construction of the catapult will begin.

7.3.1 Specific objectives:

•Stimulate creativity by building projects and carrying out practical experiments;

•Apply mathematical concepts in real-world contexts by building and programming robots;

•Develop teamwork, communication and collaboration skills by carrying out projects that combine robotics and mathematical concepts;

•Developing interest in mathematics through practical and fun activities with robotics;

•(GO-EMMAT405A) Understand the basic idea of algorithms as a finite sequence of steps (instructions), registering mathematical representations (algebraic, geometric, statistical, computational, among others) referring to everyday situations (routine or not) to organize the process and use initial programming language concepts in the implementation of algorithms written in ordinary and/or mathematical language;

•(GO-EMMAT315A) Understand the concept of a flowchart as a graphic representation of the sequence of steps in a process, reading and identifying its basic symbols (start/end, arrow, connector, among others) and types (block diagram, simple process diagram, functional diagram, horizontal diagram, vertical diagram, among others) to map information presented in situations, as well as organizing the processes, reasoning and steps for solving the problem;

•(GOEMMAT405B) Use initial programming language concepts in the implementation of algorithms written in ordinary and/or mathematical language, analyzing the results and their implications for solving problems involving algebraic expressions, functions and/or mathematical algorithms, among others.

7.3.2 Teaching resources:

•Notebook;

•Media projector or television set;

•Each group should bring to this lesson: Twelve popsicle sticks measuring approximately 12 cm, No. 120 sandpaper to finish off the cut pieces of stick, a stylus, a ruler and a barbecue skewer;

•Hot glue provided by the school;

•Chart;

•Chalk.

7.3.3 Introduction - 30 min

Teacher, review the list of materials requested in the previous lesson and talk to the students about the difficulties they encountered in acquiring the items on it. Ask if all the groups managed to get the quantity they were asked for. It's possible that some groups managed to get a larger quantity and other groups didn't get the quantity they wanted. This is the time to develop the virtues of altruism, compassion and companionship in your class. Ask the students to exchange materials if necessary.

Present the list of materials that will be used in the two lessons of this sequence and confirm that all the working groups have acquired them.

Tell the students that a very important tool for carrying out a project is to divide it into smaller tasks and propose a timetable for monitoring its execution. Present the timetable proposed in APPENDIX C - PROPOSED FLOW CHART FOR THE CONSTRUCTION OF THE CATAPULT or your own.

adaptation. Encourage the groups to divide up the tasks among their members and help them do so in order to optimize the time it takes to complete the tasks. For example: so-and-so will cut out the 5 cm pieces of toothpick, so-and-so will finish them off with 120 grit sandpaper, etc. Reinforce with the students that the description and organization of this schedule can also be called the catapult construction algorithm or flowchart.

7.3.4 Development - 50 min

Before allowing the students to start cutting, it's a good idea to go over how to use the ruler to measure the length of the toothpick. It's very common to see students who don't know how to use a ruler that's broken, as well as students who will draw a line of three centimeters starting at number 1 and going up to number 3. The lack of use of instruments for measuring length, such as rulers, measuring rods, tape measures and so on, leads to this type of error. The teacher's role is fundamental in ensuring that the measurements are cut as requested. Accompany the groups during this lesson.

It's time to get your hands dirty, or rather your stylus and ruler. Each group should make the necessary cuts in the sticks and barbecue skewer according to steps 1 and 2 of the step-by-step process for building the catapult without programming/automation. Then follow step 3 as described above.

7.3.5 Conclusion - 20 min

As a strategy for organizing the classroom for the next teacher and for monitoring the timetable, ask the students to collect the materials they have used, organize the rows and return to their seats. Present the timetable stipulated in the previous lesson and reflect with the students on the performance of each group. Ask questions such as: Is my group still on schedule? How can I help my group perform better at the next meeting? Show the students the importance of constantly evaluating the timetable and adapting it according to needs.

7.4 Lessons 05 and 06 - Catapult construction: execution of steps 4 to 8 (1h 40 min)

Lessons aimed at assembling the catapult without automation/programming. Teacher, take an active part each stage by paying attention to all the groups and encouraging them. This is the time to get closer to the students and create a bond of trust and partnership that is necessary for a good development of the theoretical lessons to come. Use the laptop and media projector to show the students images of what the catapult should look like at the end of each step.

7.4.1 Specific objectives:

- Stimulate creativity by building projects and carrying out practical experiments;
- Applying mathematical concepts in real-world contexts by building and programming robots .
- Develop teamwork, communication and collaboration skills by carrying out projects that combine robotics and mathematical concepts;
- Developing interest in mathematics through practical and fun activities with robotics;
- (GO-EMMAT405A) Understand the basic idea of algorithms as a finite sequence of steps (instructions), registering mathematical representations (algebraic, geometric, statistical, computational, among others) referring to everyday situations (routine or not) to organize the process and use initial programming language concepts

in the implementation of algorithms written in ordinary and/or mathematical language;

7.4.2 Teaching resources:

- Notebook;
- Media projector or television set;
- Each group should bring to this lesson: Twelve popsicle sticks measuring approximately 12 cm, No. 120 sandpaper to finish off the cut pieces of stick, a stylus, a ruler and a barbecue skewer;
- Hot glue provided by the school;
- Chart;
- Chalk.

7.4.3 Introduction - 10 min

Take the students back to the activity carried out in the previous lesson and present the updated timetable again to keep the time organized and the students engaged.

7.4.4 Development - 70 min

This lesson will cover steps 4 to 8 of the step-by-step construction of the catapult without automation/programming. Always accompany the students during this stage. The time taken to assemble the catapult will depend on the number of hot glue guns available to the students. If each group has a gun, these assembly lessons will take less time.

7.4.5 Conclusion - 20 min

At the end of step 8, the catapult is ready to launch objects and projectiles. I'm sure the students will test the launches without the teacher asking them to. Let the students enjoy the fun and take the opportunity to ask the groups new questions during the launches. For example: What affects the distance reached by the projectile? Expected answers: the mass of the projectile, the size of the projectile, the shape of the projectile, the amount I stretch the rubber band for the launch.

7.5 Lessons 07 and 08 - Finalizing the catapult: execution of step 9 and initial launch tests (1h 40min)

7.5.1 Specific objectives:

- Apply mathematical concepts in real-world contexts by building and programming robots;
- Study the influence of variables such as launch angle, object mass and applied force on the distance traveled by the launched object, promoting an understanding of the physical principles involved in movement;
- Develop teamwork, communication and collaboration skills by carrying out projects that combine robotics and mathematical concepts;
- Developing interest in mathematics through practical and fun activities with robotics;
- Understand basic concepts of robotics and programming;
- Carry out experiments to collect data on the launches made by the catapult, making it possible to analyze graphs of position, velocity and acceleration a function of time;
- Relate the angle of the catapult shot in oblique launches to establish the maximum range of the projectile;
- (GO-EMMAT405A) Understand the basic idea of algorithms as a finite sequence of steps (instructions), registering mathematical representations (algebraic, geometric, statistical, computational, among others) referring to everyday situations (routine or not) to organize the process and use initial programming language concepts in the implementation of algorithms written in ordinary and/or mathematical language;

7.5.2 Teaching resources:

- Notebook;
- Media projector or television set;
- Hot glue provided by the school;
- Ruler;
- 180° protractor;
- Board protractor;
- Stylus;
- Two 5m long tape measures;
- 2.5m long string;
- Chart;
- Chalk.

7.5.3 Introduction - 20 min

Teacher, this is the lesson in which the catapult will be ready to launch projectiles with different launch angles. Remind the students of the questions asked in the previous lesson about which variables affect the projectile's range. The angle of launch has probably not yet been hypothesized. This is the time to raise this question before cutting out and gluing the cardboard to the catapult. Ask: If the catapult launches the projectile before it hits the launch limiter, will the range be longer or shorter? Listen to the students' answers. There is unlikely to be unanimity. It's time to explain that the function of this last step is to force the catapult to launch the projectile at different angles precisely in order to carry out tests and observe which launch angle causes the projectile to reach the greatest horizontal distance.

7.5.4 Development - 60 min

Many students finish high school without ever using a protractor in a practical lesson. So introduce the protractor as a tool for measuring angles. may also need to remind the students what an angle is. Use the protractor board, if available in your school, and instruct the students on how to construct and measure angles using a ruler and protractor. It's preferable to instruct students and the school to buy 180° protractors because it's "two in one". It works as both a ruler and a protractor. Construct the notable angles of 30°, 45° and 60° on the board. Emphasize the elements of the angle: origin, vertex and side. Ask the students to construct these angles in their notebooks to get used to using the protractor. Walk around the classroom to follow along with this activity.

With the students sharp in using the protractor and constructing the angles, guide them to cut the cardboard according to step 9 of constructing the catapult without automation/programming.

To eliminate the variables such as the mass and shape of the projectile and to make it easier to mark the distance reached, it is advisable to carry projectiles made from bags, sand and string, like the one in the photo below.

Figure 58 - Projectile made from a plastic bag, sand and string

Source: The author

Other projectiles were used, such as marbles, school erasers and stones. However, they had difficulty identifying the distance reached for measurement. This improvisation with a small bag, sand and string was considered because the sand helps to cushion the fall of projectile after launch and makes it easier to identify the distance reached.

To make this projectile model, simply use a soft drink cap as a measure of sand and stick it inside a plastic bag. Then tie the bag very close to the mound formed by the sand and cut off the excess bag and string. The soda cap should be used as a unit of measurement so that the projectile fits perfectly into the throwing arm of the catapult. Another positive point of making this projectile is to standardize at least this element before writing down the measurements achieved by its launch.

To record the distances reached in the table below, organize the room so that there is a free corridor through which the two 5m long measuring rods are opened parallel to each other. At each launch, the catapult must be positioned at the zero point of the tape measure.

The dynamics of the launches and recording the distances will be as follows: at each launch, two students from the group will stand on the catapult to hold it and trigger it manually. When the projectile is launched, the other two students hold the 2.5m string by the end and position it over the same length of each of the tape measures on top of where the projectile stopped, regardless of whether it rolled or not. The students who triggered the catapult write down the measurement in the table below.

Table 1 - Table recording the distances reached by the projectile at different launch angles

Launch angle	Launches					Average distance per throw (in cm)
	1º	2º	3º	4º	5º	
30°						
45°						
60°						
90°						

Source: The author

The table below shows the actual measurements achieved by the catapult presented in the general guidelines for this didactic sequence.

Table 2 - Completed distance recording chart

Launch angle	Launches					Average distance per throw (in cm)
	1º	2º	3º	4º	5º	
30°	61	69	53	57	60	60
45°	113	113	116	113	113	113,6
60°	204	200	207	190	201	200
90°	227	243	277	265	244	251,2

Source: The author

Wait until all the groups have made their respective entries and filled in the distance chart. The time needed for this stage may vary from one class to another, depending on organization and discipline.

7.5.5 Conclusion - 20 min

Time for each group to discuss the results recorded and answer some questions such as: Which launch angle had the greatest average distance achieved? Which launch angle had the shortest average distance achieved? Does the measurement recorded correspond to the first point where the projectile reached the ground? Was the initial launch velocity the same for each launch angle? Why?

Do not provide the information that, under ideal conditions, the angle of 45° is the one with the greatest horizontal distance achieved.

7.6 Lessons 09 and 10 - Oblique launch: real vs. ideal, exploring a digital simulation (1h 40min)

Teacher, it's time to start relating the practical activity carried out by the students to the physical and mathematical concepts involved. However, it is recommended that this transition from the practical to the conceptual be conducted smoothly in order to keep the students engaged in the proposed activities.

7.6.1 Specific objectives:

- Study the influence of variables such as launch angle, object mass and applied force on the distance traveled by the launched object, promoting an understanding of the physical principles involved in movement;
- Promote the development of logical thinking and problem solving through the application of mathematical concepts in robot programming and control;
- Developing interest in mathematics through practical and fun activities with robotics;
- Relate the angle of the catapult shot in oblique launches to establish the maximum range of the projectile;
- Encourage a interdisciplinarity, integrating objects of knowledge of mathematics, physics and science and technology through projects involving robotics;

7.6.2 Teaching resources:

- Notebook;
- Media projector or television set;
- Internet;
- Mobile lab with Chromebook;
- Chart;
- Chalk.

7.6.3 Introduction - 20 min

Review the results obtained in the previous lesson with the groups. Ask each group which angle achieved the greatest horizontal reach and discuss once again which variables influence the distance reached by the projectile. At this point, the teacher needs to lead the discussion to the object of knowledge of the lesson: oblique launch. Make a note on the board of the main variables involved: the mass of the projectile, the launch angle and the initial launch velocity.

7.6.4 Development - 60 min

Invite students to access the interactive simulation "Projectile movement" on the *Phet Interactive Simulations* website via the link provided in the support material for this lesson. This site offers various interactive simulations for studying, developing and visualizing concepts in mathematics, chemistry, physics, biology and earth sciences.

Tell the students that the simulation they will be using allows them to easily modify the variables that have been noted on the board and identified as those that affect the horizontal range reached by the projectile. You and the students will notice that the simulation is very intuitive and easy to use. You can change the type of projectile to be launched by the cannon: cannonball, golf ball, pumpkin, human, piano, car, among others. The launch height, initial velocity and launch angle, as well as air resistance, are all variables that can be changed in this simulation.

Make several shots with the students, changing all the possible variables so they can observe. If your school has a computer lab and each student or group students has a computer connected to the internet, encourage them to also take shots and explore the simulation until they are used to handling it.

Ask the students to reproduce the tests carried out in class with the catapult in this simulation. Tell them to copy the table recording the distances with the notable angles into their notebooks and write down the new results. To standardize the results, set the initial launch height at 5m, the initial speed at 15m/s and the type of projectile as a cannonball. You can measure the horizontal distance reached by the projectile by positioning the rectangle with the time, range and height information on top of the projectile, as shown in the image below.

Figure 59 - Rectangle positioned in the Projectile Movement simulation on the *Phet Interactive Simulations website* to obtain the horizontal distance reached by the projectile

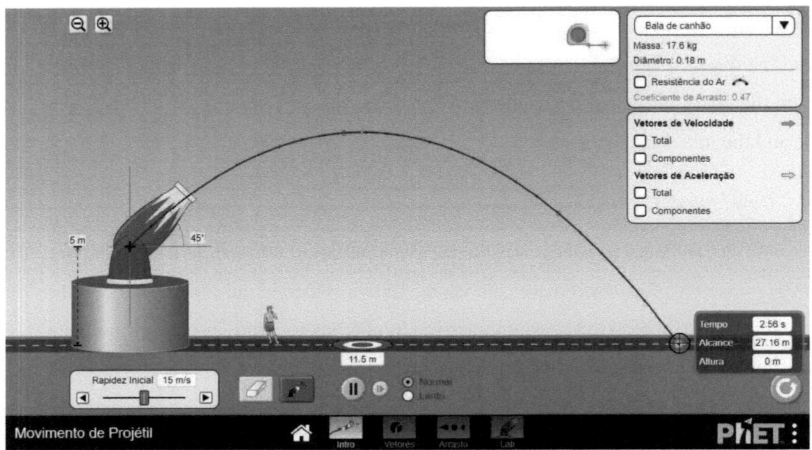

Source: Available at: https://phet.colorado.edu/sims/html/projectile-motion/latest/projectile-motion_all.html?locale=en_BR. Accessed on: July 4, 2024.

After all the entries have been made, the board for each student or group of students should show exactly the results.

Table 3 - Table recording the distance reached by the cannonball in the simulation Projectile movement with height= 5m, initial speed = 15 m/s

Launch angle	Horizontal distance reached (in m)
0°	15,14
30°	26,38
45°	27,16
60°	22,42
90°	0

Source: The author

The students probably won't add the zero angle to the tally chart. Explain that the angle corresponding to the 90° launch angle on the catapult is the zero angle in the interactive simulation and guide them to add it and make the launch.

7.6.5 Conclusion - 20 min

Discuss the results obtained in both tests: the real test carried out with the catapult and the ideal test carried out in the simulation. Ask the following questions: In the simulation, which launch angle had the greatest distance achieved? Which launch angle had the shortest distance achieved? Does the measurement recorded correspond to the first point where the projectile reached the ground? Was the initial launch speed the same for each launch angle? Why is it that in the tests carried out with the catapult, it wasn't the 45° angle that achieved the greatest distance? Ask the students to record their answers.

At this point, students need to remember that when obtaining the measurements achieved by the catapult projectile, it was not possible to consider the first point at which it touched the ground, but its final position. This factor is added to all the other variables already discussed and especially the initial launch velocity (which is different for each catapult launch angle) while in the simulation it is possible to guarantee the same for the various angles.

7.6.6 Suggested support materials

PhET Interactive Simulations, 2024. Interactive Simulations for Science and Mathematics. Available at: <https://phet.colorado.edu/pt_BR/simulations/projectile-motion>. Accessed on: July 4, 2024.

7.7 Lessons 11 and 12 - Introduction to programming in *TinkerCad* - automatic operation of a servo motor (1h 40 min)

Professor, these two lessons are designed to introduce and familiarize students with the TinkerCad simulation platform and one of the electronic components that will be used to automate the catapult: the servo motor.

7.7.1 Specific objectives:

- Get to know the *TinkerCad* simulation platform;
- Learn the basic programming of a servo motor;
- Understand what Arduino is and its application;
- Learn the basics of programming;

7.7.2 Teaching resources:

- Notebook;
- Media projector or television set;

•Internet;

•Mobile lab with Chromebook;

•Chart;

•Chalk.

7.7.3 Introduction - 40 min

Before introducing the TinkerCad platform to the students, it is advisable to remind them of some basic electronic concepts such as electric current, resistance and potential difference. Relax! It's really just the basics. To do this, go to the *Phet Interactive Simulations* website and search for the interactive simulation "DC circuit assembly kit". This simulation is excellent because it illustrates the movement of electrons inside a conductive wire and their effect when they pass through an electronic component.

In the simulation, which is also very intuitive, add a battery, a light bulb, a switch and the wires needed to connect a simple electrical circuit like the one in the image below.

Figure 60 - Circuit with open switch in the simulation DC Circuit Assembly Kit

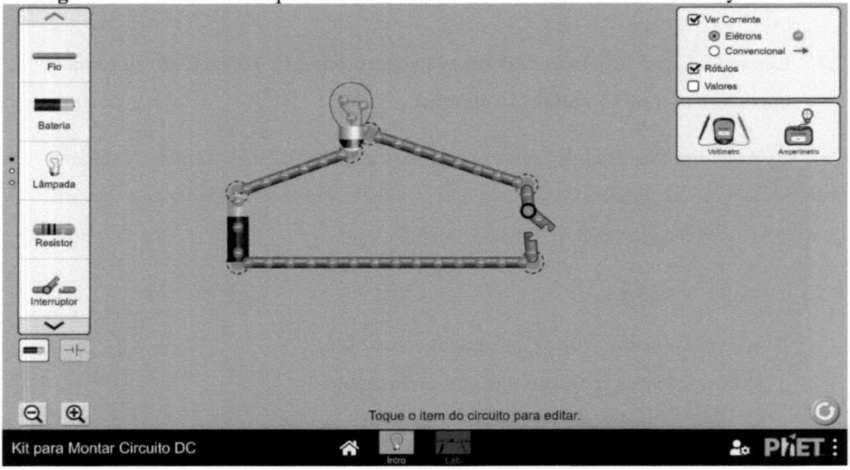

Source: Available at: https://phet.colorado.edu/sims/html/circuit-construction-kit-dc/latest/circuit-construction- kit-dc_all.html?locale=en_BR. Accessed on: 05 Jul 2024.

Ask the students to repeat the same steps and then flick the switch to close the circuit as in the image below.

Figure 61 - Circuit with open switch in the simulation DC Circuit Assembly Kit

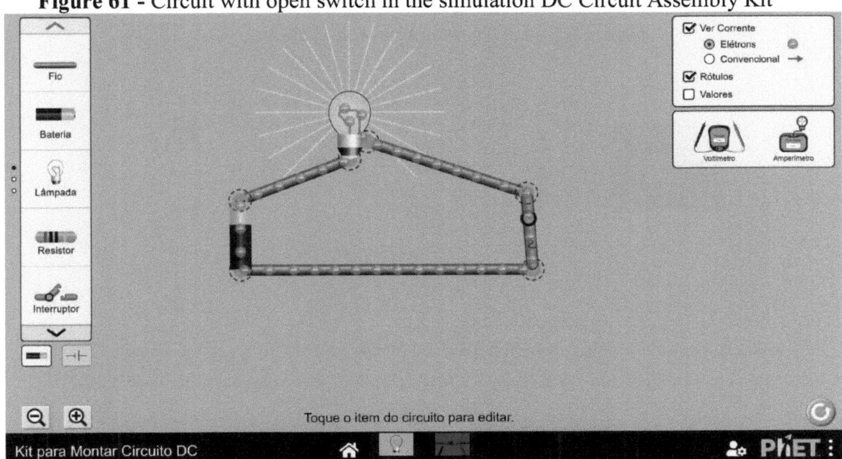

Source: Available at: https://phet.colorado.edu/sims/html/circuit-construction-kit-dc/latest/circuit-construction- kit-dc_all.html?locale=en_BR. Accessed on: 05 Jul 2024.

Take advantage of the simulation to explain that the light bulb lit up as an effect of the electric current that passed through it (the electric current is symbolized in the simulation by the movement of the little balls with a minus sign). The electric current, in turn, only existed because of a potential difference caused by the battery used in the circuit and the conductive wires that allowed the free electrons to move. Point out that these are the basic elements of an electric circuit: a source that generates a potential difference (in this case the battery), the electric current (movement of the electrons) and the conductor (the wire).

Now introduce the TinkerCad platform. Explain that this platform is used, among other things, for circuit simulations and Arduino programming and instruct the students to create a personal account so that the projects they create are saved and make easier to return to the activities in later lessons.

Reproduce in TinkerCad the same simple circuit that was created on the other site to introduce some basic platform tools and reinforce basic electronics concepts. Add a light bulb, a 9V battery, a switch and the conductive wires. Tell the students that, as a standard, red wires will always be used for the positive and black wires for the negative.

The circuit should look like the following illustration.

Figure 62 - Simple circuit built in *Tinkercad*

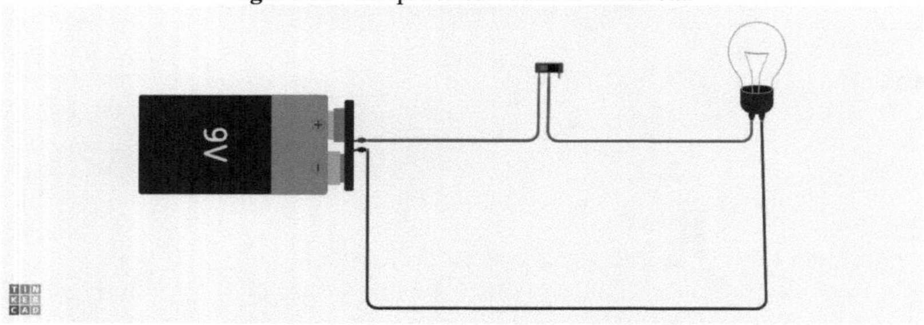

Source: Available at: https://www.tinkercad.com/things/5d8SbzXmJWV-laboratorio-de-exploracao/editel?returnTo=%2Fthings%2F5d8SbzXmJWV-motor-de-passo. Accessed on: 05 Jul 2024.

So far the aim has been to emphasize to the students what is needed for there to be an electric current so that the components connected to the circuit work correctly. Demonstrate that if the circuit isn't closed, there won't be any current either, let alone its effect. Replace the light bulb with a DC (Direct Current) motor to illustrate that the effect electric current can also be different in different electronic components. In the light bulb the current makes it light up, in the motor it makes it turn. Remember that you have to click on "Start simulation" for the circuit to start working.

7.7.4 Development - 40 min

Create a circuit called Servo motor controlled by Arduino and ask the students to do the same, each on their own. Add the Arduino Uno board to the circuit. Explain that the added board has some important functions such as: a 5V power supply for the circuit when connected to a computer via a USB (*Universal Serial Bus*) cable or a power supply of between 7 and 12 V (recommended voltages to avoid overheating) when connected to an external power supply; a microcontroller that allows you to program commands that can be executed automatically or via actuators such as buttons, for example. Remember that this is a lesson in basic concepts. Don't go into too much depth.

Add a small test board to the circuit, also known as a *protoboard*. The students should repeat the steps with your guidance. Explain that this board has the function of connecting the elements of a circuit without the need to solder them. It acts as a bridge through which electric current is carried to different places (components). Connect the Arduino board to the test board by adding conductive wires from the 5V pin to the red line on the test board. Make it clear to the students that this will cause electric current to flow through the test board when the Arduino is connected via USB to a computer or an external source.

Add a DC (Direct Current) motor. Explain that this motor can easily be found in toys such as remote control cars, toy cars that go "by themselves" after you put in a battery and flip a switch. Make the connections as shown in the illustration below, then click on "Start simulation" to see the motor rotating and explain that this is because the electric current supplied by the computer passes to the Arduino board via the USB cable, is conducted to the protoboard via the conductive wires and, finally, is transported to the motor also conducted by the conductive wires. These basic concepts are fundamental for proper handling and connection of the circuits built with the Arduino board and protoboard. Reinforce with the students that, in practice, it can happen that the conductor wires are not well connected and the current does not flow properly.

Figure 63 - Simple circuit with DC motor, Arduino board and protoboard

Source: Available at: https://www.tinkercad.com/things/dY2uxz9aBC3-dazzling-tumelo-elzing/editel?tenant=circuits. Accessed: 06 Jul 2024

Now replace the DC motor with a micro servo and make the connections as shown in the illustration

next.

Figure 64 - Simple servo motor circuit on protoboard and Arduino Uno

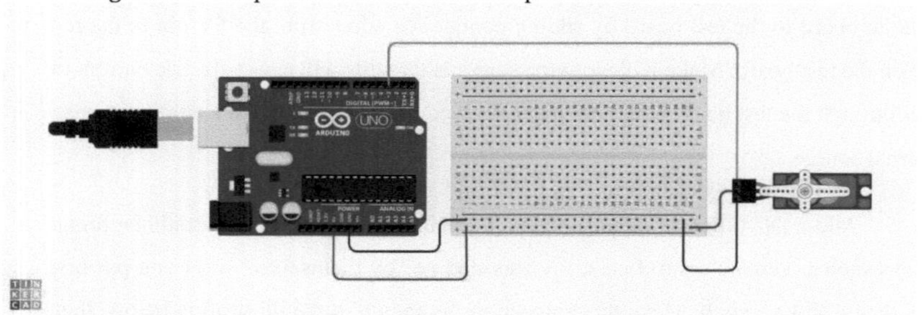

Source: Available at: https://www.tinkercad.com/things/dY2uxz9aBC3-dazzling-tumelo-elzing/editel?tenant=circuits. Accessed: 06 Jul 2024

It is to be hoped that, because many students are curious and take the initiative, they will click on "start simulation" before you ask them to. If don't, ask them to click on this function and wait for their reaction. It's important that the teacher doesn't explain everything at once and lets curiosity and unexpected reactions happen during the lesson. It's a strategy to keep the class engaged and attentive. Explain that this motor doesn't "work" like the DC motor because it needs a command to work. That's why it has a third wire (the orange one).

The orange wire is responsible for receiving information from the Arduino and running the micro servo accordingly. This wire must always be connected to one of the Arduino's PWM (*Pulse Width Modulation*) ports. Also known as analog ports, the PWM ports on an Arduino UNO board are those with the uncle sign before the number (~3, ~5, ~6, ~9, ~10 and ~11). Basically, they allow you to set different voltage values while the digital ports only allow two: full voltage or zero voltage. In other words, the digital ports (2, 4, 7, 8, 12 and 13) either turn a motor on or off and the analog ports allow you to start the motor and make it rotate at different speeds depending on the voltage configured.

Explain to the students that this information is passed to the micro servo via programming codes. To do this, click on "code" in *TinkerCad* and select text mode. By default, *TinkerCad* always offers a code ready to turn an LED on for one second and off for one second as well. Delete the code *TinkerCad* displays and paste the code shown below.

Figure 65 - Code for basic programming of a servo motor

```
//Código para programação básica de um servo motor
//Autor: Fausto José Fernandes Dias
//Data: 11/07/2024

#include <Servo.h> // inclui a biblioteca do servo motor

int PosIniServo = 0;// define a posição inicial do servo
                    //como zero grau

Servo SM;// cria um objeto para controlar o servo motor (SM);

void setup()
{
  SM.attach(3);// informa ao Arduino qual a porta enviará
               //informações ao servo motor
}

void loop()
{
  // Faz o servo motor girar de 0 a 180 graus de grau em grau
  for (PosIniServo = 0; PosIniServo <= 180; PosIniServo += 1) {
    // Lê a posição e informa ao servo motor
    SM.write(PosIniServo);
    // Faz o servo aguardar um décimo de segundo antes de passar
    // para a próxima posição
    delay(100);
  }
}
```

Source: The author

7.7.5 *Conclusion - 20 min]*

Promote a moment of experimentation in your class and encourage students to change information in the code, such as the analog port, the number of degrees the servo must move in turn, and the waiting time to move to the next position. A very common mistake both when copying the code and when manipulating it is not paying attention to the words written in uppercase and lowercase letters. For example, "PosIniServo" is different from "Posiniservo". This means that if you don't write all the code with the same writing pattern, the program won't understand what you've set it up to do.

7.7.6 *Suggested support materials*

PhET Interactive Simulations, 2024. Interactive Simulations for Science and Mathematics. Available at:< https://phet.colorado.edu/pt_BR/simulations/circuit-construction-kit-dc>. Accessed on: July 5, 2024.

7.8 Lessons 13 and 14 - Servo motor controlled by a button (1h40min)

In these two lessons, the code written in the previous lessons will be improved and a *push button* will be added to drive the servo motor. Students will then assemble the circuit in practice to see how it works. If your school doesn't have enough kit or doesn't have any kit, you can do the programming only on the TinkerCad platform and then select some students to assemble the circuit in practice under guidance.

7.8.1 Specific objectives:

- Understand what Arduino is and its application;
- Familiarize yourself with the Arduino IDE;
- Learn the basics of programming;
- Developing interest in mathematics through practical and fun activities with robotics;
- Understand basic concepts of robotics and programming;
- To arouse interest in the study of robotics.

7.8.2 Teaching resources:

- Notebook;
- Media projector or television set;
- Internet;
- Mobile lab with Chromebook;
- Arduino kit;
- Chart;
- Chalk.

7.8.3 Introduction - 30 min

Tell the students that the code written in the previous lesson will be improved and a new component added: a button. With the *TinkerCad* page open, ask the students to create a copy of the file from the previous lesson and rename the copy "Servo motor driven by a button". Search for button in the *TinkerCad* search bar and add a button to the circuit as shown in the following image.

Figure 66 - Circuit of a button-operated servo motor in *TinkerCad*

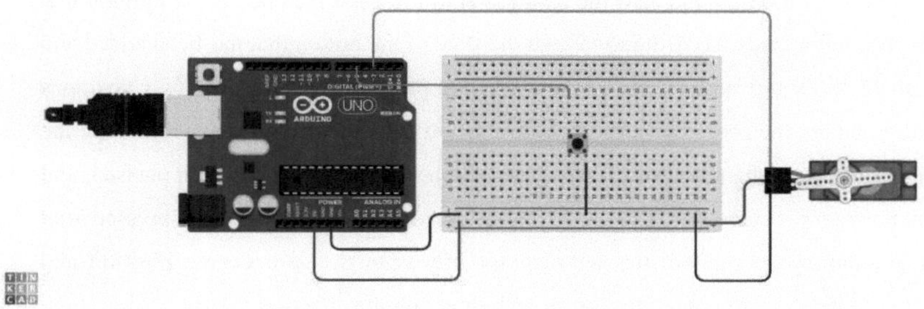

Source: Available at: https://www.tinkercad.com/things/f8cg3gYYUbg-servo-motor-acionado-por-um- botao/editel. Accessed on: July 11, 2024.

Then instruct them to make the changes to the code as shown in the image. Emphasize that the servo motor will be the component that fires the catapult.

Figure 67 - Code to control a servo motor via a button

```
//Código para controlar um servo motor através de um botão
//Autor: Fausto José Fernandes Dias
//Data: 11/07/2024

//Inclui a biblioteca do servo motor
#include <Servo.h>

Servo SM; // Cria um objeto para controlar o servo motor (SM)
int BotaoSM = 5; // o botão que aciona o SM será ligado na porta 5 do arduíno
int Servo = 3; // SM ligado na porta 3 do arduíno
int PosIniServo = 0; // Posição inicial do servo no ângulo zero
bool BotaoPressionado = false; // Variável para rastrear o estado do botão

void setup() {
  SM.attach(Servo); // Conecta o servo ao pino
  pinMode(BotaoSM, INPUT_PULLUP); // Configura o botão como entrada com pull-up interno
}

void loop() {
    //Informando o que deve acontecer se o botão for pressionado
  if (digitalRead(BotaoSM) == LOW){ // Se o botão for pressionado
    if (!BotaoPressionado){// Verifica se o botão estava solto na iteração anterior
      PosIniServo = 90; // Define a posição do servo (por exemplo, 90 graus)
      SM.write(PosIniServo);// Move o servo para a posição desejada
      BotaoPressionado = true; // Marca o botão como pressionado
      delay(200); // pequeno atraso para dar tempo de realizar a leitura
    } else { // Se o botão já estava pressionado
      PosIniServo = 0; // // Volta à posição inicial (0 grau)
      SM.write(PosIniServo);
      BotaoPressionado = false; // Marca o botão como solto novamente
      delay(200);
    }
  }
}
```

Source: The author

As with the previous code, this is a code that presents more advanced functions with the "*if...else*" conditional function. Once , just ask the students to make a copy of the code in *TinkerCad* and start the simulation to see what the code does. Explain that this code commands the Arduino to read the state of the button that has been added and to move the servo motor only if the button is pressed. And that the two positions configured are the zero degree position and the 90 degree position. This way, when the simulation starts, the program will read the button, which has not yet been pressed, and then position the servo motor in the 90 degree position. When the button is pressed after the simulation has started, the servo motor moves to the zero degree position and alternates between zero and 90 degrees each time button is pressed.

7.8.4 Development - 50 min

If you've managed to get your school to buy an Arduino kit, now is the time get the students to repeat this experiment. Introduce and explain the Arduino board, the test board and the servo motor once again, with each student handling these components. Explain that the code that the Arduino board will receive needs to be typed into a computer application: the Arduino IDE. Instruct the students to download this application on the mobile lab's Chromebook and copy the code from *TinkerCad* and paste it into the Arduino IDE.

As is a practical activity, it requires constant attention and monitoring by the teacher in each group. If your school hasn't bought enough kits for groups of four students, you could use just one kit and select a few students from each group to connect the jumpers and the circuit, as well as the button tests.

7.8.5 Conclusion - 20 min

Reinforce with the students that the *TinkerCad* platform is very important because it allows them to test not only the code that has been written but also the connections that have been made without the need to own the electronic components or the Arduino kit. Please note that it is always recommended to run the simulation before starting to connect the devices to the socket or Arduino board to prevent them from being burnt out by excess current. Make a simulation with an LED in *TinkerCad* to demonstrate that it is very easy to lose physical electronic components by just plugging them in to investigate how they work.

7.8.6 Suggested support materials:

Autodesk TinkerCad, 2024. LED driven by a button. Available in:

<https://www.tinkercad.com/things/ayQ1WvX4qXW-led-acionado-por-um-botao/editel>. Accessed on: July 11, 2024.

7.9 Lessons 15 and 16 - Stepper motor controlled by a button (1h 40min)

Unlike the lessons on the servo motor, where programming began with the *TinkerCad* simulator, for programming the stepper motor it is advisable to start the assembly in practice and run the program in the Arduino IDE. This is because the simulator requires additional components that are slightly different from those that will be used in practice. In practice, the ULN2003 stepper motor *driver* has a much more intuitive connection than the possible simulations in *TinkerCad.*

7.9.1 Specific objectives:

- Understand what Arduino is and its application;
- Familiarize yourself with the Arduino IDE;
- Learn the basics of programming;
- Developing interest in mathematics through practical and fun activities with robotics;
- Understand basic concepts of robotics and programming;
- To arouse interest in the study of robotics.

7.9.2 Teaching resources:

- Notebook;
- Media projector or television set;
- Internet;
- Mobile lab with Chromebook;
- Arduino kit;
- Chart;
- Chalk.

7.9.3 Introduction - 30 min

Teacher, introduce the students to the new components that will be used in this lesson: the stepper motor and the ULN2003 *driver*. Explain that the stepper motor, unlike the servo motor, can rotate 360 degrees and can also make several clockwise and anti-clockwise turns. In , the stepper motor has much more "force" than the servo motor due to the gear system inside it. The ULN2003 driver will be used to allow the stepper motor to receive the commands that will be entered into the Arduino IDE

Ask the students to connect the circuit as shown in the diagram below.

Figure 68 - Circuit diagram of the stepper motor driven by a button

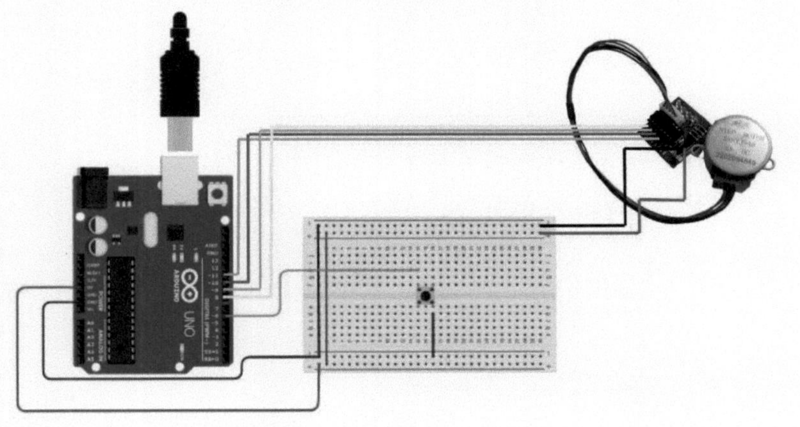

Source: The author

7.9.4 Development - 50 min

Next, ask them to copy the code shown in the figure below, which will allow you to control the stepper motor using the button.

Figure 69 - Code to drive a stepper motor using a button

```
//Código para acionar um motor de passo através de botão
//Autor: Fausto José Fernandes Dias
//Data: 15/07/2024

#include <Stepper.h>
#define e1 8
#define e2 9
#define e1 10
#define e2 11

// Define o número de passos por revolução do motor
const int passosPorRevolucao = 64;

// Configuração do motor de passo
Stepper motor(passosPorRevolucao,11, 10, 9, 8);

// Pino do botão
const int pinoBotao = 6;
int estadoBotao = HIGH;
int ultimoEstadoBotao = HIGH;
bool sentidoHorario = true; // Inicialmente, gira no sentido horário

void setup() {
  // Inicializa o motor
  motor.setSpeed(100); // Velocidade em RPM (ajuste conforme necessário)
  pinMode(pinoBotao, INPUT_PULLUP);
}

void loop() {
  // Lê o estado do botão
  estadoBotao = digitalRead(pinoBotao);

  // Verifica se o botão foi pressionado
  if (estadoBotao == LOW && ultimoEstadoBotao == HIGH) {
    // Alterna o sentido do motor
    sentidoHorario = !sentidoHorario;

    // Gira o motor continuamente no sentido selecionado
    while (digitalRead(pinoBotao) == LOW) {
      if (sentidoHorario) {
        motor.step(1); // Um passo no sentido horário
      } else {
        motor.step(-1); // Um passo no sentido anti-horário
      }
      delay(10); // Pequeno atraso para evitar sobrecarga
    }
  }
  // Atualiza o estado anterior do botão
  ultimoEstadoBotao = estadoBotao;
}
```

Source: The author

Clarify that the copied code causes the stepper motor to change its operation whenever the button is pressed continuously. If the button is held down, the stepper motor rotates clockwise. Once released and pressed again, the stepper motor turns counterclockwise.

7.9.5 Conclusion - 20 min

Finally, tell the students that in the next lesson all these codes and circuits will be integrated with the catapult in order to automate it, i.e. make it trigger and fire using programs commands sent via these circuits.

7.10 Lessons 17 and 18 - Catapult program in Arduino for the study of oblique launching in mathematics

Finally, these lessons will integrate the catapult already built and tested by the students with the circuits and electronic components presented in the previous lessons.

7.10.1 Specific objectives:

- Developing a catapult controlled Arduino programming;
- Stimulate creativity by building projects and carrying out practical experiments;
- Program the catapult to launch at different launch angles;
- Carry out experiments to collect data on the launches made by the catapult, making it possible to analyze graphs of position, velocity and acceleration a function of time;
- Relate the angle of the catapult shot in oblique launches to establish the maximum range of the projectile;

7.10.2 Teaching resources:

- Notebook;
- Media projector or television set;
- Internet;
- Mobile lab with Chromebook;
- Arduino kit;
- Hot glue gun;
- 30 cm piece of wood (*pallet*);
- Chart;
- Chalk.

7.10.3 Introduction - 20 cm

Start this lesson by projecting an image of the automated catapult so that the students have a good idea of how it should look. Instruct them that it is very important to position the servo motor correctly so that it is able to "lock" the throwing arm correctly. Another precaution that needs to be observed is the position of the stepper motor so that it doesn't get in the way of the total traction of the throwing arm. Make the students aware that the correct positioning and marking of this position is very important.

7.10.4 Development - 40 min

Repeat with the students steps 1 to 4 of Step-by-step programming and automation of the catapult presented in the general guidelines for this teaching sequence. Accompany the students through each step and, if necessary, use photographic records of the step-by-step to help them.

7.10.5 Conclusion - 30 min

Once the catapult has been assembled and programmed, ask the groups to make new launches with the catapult and record the results on the table of launches achieved and compare the results obtained without the automation with the results obtained with the automation. It is likely that the rubber band used prior to automation will have to be "relaxed" so that the stepper motor has enough force to drive the catapult. If this happens, the results obtained will probably be lower than those previously recorded.

8 CONCLUSION

In view of the , building a catapult is a process that encourages the development of basic concepts in robotics, physics and mathematics: planning, organization and creativity. The relationship between these areas has yet to be addressed in secondary school in a superficial way due to the level of knowledge of students at this stage of education. A different and innovative strategy such as the one presented contributes to student engagement and motivation, improves the relationship with the teacher and, consequently, enhances the development of the competences and skills proposed by the BNCC.

Therefore, the practical experiments presented also help students to clearly understand the related concepts and objects of knowledge: the oblique launch, the trajectory of the projectile and its relationship with the parabola. Concepts which, in many cases, were worked on in the classroom abstractly and mechanically in order to limit themselves to handling formulas and substituting certain values to analyze the effects.

9 BIBLIOGRAPHICAL REFERENCES

BRAZIL. **Common National Curriculum Base**. Brasília: Ministry of Education,

2018. CAMPOS, F. R. EDUCATIONAL ROBOTICS IN BRAZIL: OPEN

QUESTIONS,
CHALLENGES AND FUTURE PERSPECTIVES. **Ibero-American Journal of Studies in Education**, Araraquara, Oct. 15, 2017. 2108-2121.

CARVALHO, N. A. D.; AGUIAR, D. S. D. AUTOMATED CATAPULT FOR TEACHING OBLIQUE LAUNCH. **Proceedings of the National Robotics Exhibition - MNR 2018**, 2018. 87-89.

GARCIA, G. R.; LABRE, T. H. M. The pedagogical challenge of the alpha generation. **Culturas & Fronteiras Magazine**, Porto Velho - RO, December 2021. 39-58.

GOIÁS. **Curriculum Document for Goiás - High School Stage**. Goiânia: State Department of Education, 2020.

MEDEIROS, D. G. D. **The warlike use of science: the example of catapults as an application of theories of elasticity and oblique launch**, Niterói, 2014.

UNIVERSITY OF COLORADO BOULDER. PhET Interactive Simulations. **PheT Colorado**, 2024. Available at: <https://phet.colorado.edu/pt_BR/simulations/projectile-motion>. Accessed on: March 03, 2024.

APPENDIX A - CODE IN ARDUIIN FROM PROGRAMMING AND AUTOMATION OF THE CATAPULT

```
//FEDERAL UNIVERSITY OF CATALÃO
//INSTITUTE OF MATHEMATICS AND TECHNOLOGY
//SPECIALIZATION COURSE IN EDUCATIONAL ROBOTICS AND ITS TECHNOLOGIES IN
THE TEACHING OF MATHEMATICS
//Catapult programmed in Arduino to study oblique launch in
mathematics
//Author: Fausto José Fernandes Dias
//Supervisor: Prof. Dr. Flávia Gonçalves Fernandes Dr. Flávia Gonçalves
Fernandes

#include <Servo.h> // includes the servo motor library
#include <Stepper.h> // includes the stepper motor
library
#define e1 8 //define that input 1 of the stepper motor driver will be
connected to port 8 of the arduino
#define e2 9 //define that input 1 of the stepper motor driver will be
connected to port 9 of the arduino
#define e1 10 //define that input 1 of the stepper motor driver will be
connected to port 10 of the arduino
#define e2 11 //define that input 1 of the stepper motor driver will be
connected to port 11 of the arduino

// Stepper Motor Settings (MP)
const int stepsTurn= 64; // each MP has a number of steps per turn, the
type 28BYJ-48 (5V) used in this project is 64
Stepper motor(stepsByTurn, 8, 10, 9, 11); // set the order in which each
coil in the stepper motor will be actuated, note that the second coil
must be interchanged with the third.
const int BotaoMP=  6; // the button that drives the Stepper Motor will
be connected to port 6 of the arduino
int statusBotaoMP= HIGH; //initializes the button as HIGH (on)
int lastButtonState= HIGH; //initializes the last button state as HIGH
(on)
bool clockwise= true; // Initially turns clockwise

//Servo Motor Settings (SM)

Servo SM; // Creates an object to control the servo motor (SM)
int BotaoSM= 5; // the button that triggers the SM will be connected to
port 5 of the arduino int Servo= 3; // SM connected to port 3 of the
arduino
int PosIniServo= 0; // Initial position of the servo at zero angle
bool ButtonPressed= false; // Variable to track the state of the button

void setup() {
  motor.setSpeed(900); // Speed in RPM (adjust as necessary)
```

```
  pinMode(BotaoMP, INPUT_PULLUP);
  SM.attach(Servo); // Connect the servo to the pin

  pinMode(BotaoSM, INPUT_PULLUP); // Set the button as an internal
pull-up input
}

void loop() {
  //No Stepper motor
    statusBotaoMP= digitalRead(BotaoMP);// Read button status

  if (stateBotaoMP == LOW && lastStateBotaoMP == HIGH) {// Check if
the button has been pressed
    // Toggle the motor direction
    directionTime= !directionTime;

    // Turns the motor continuously in the
    selected direction while (digitalRead(BotaoMP)
    == LOW) {
      if (directionTime) {
        motor.step(1); // One step clockwise
      } else {
        motor.step(-1); // One step counterclockwise
      }
      delay(10); // Short delay to avoid overloading
    }
  }
  // Update the previous state of
  the lastBotaoMPState button=
  BotaoMPState;

  //No Servo Motor
  if (digitalRead(BotaoSM)== LOW){ // If the button is pressed
    if (!ButtonPressed){// Checks if the button was released in the
previous iteration
      PosIniServo = 90; // Set the servo position (e.g. 90 degrees)
      SM.write(PosIniServo);// Move the servo to the desired position
      ButtonPressed= true; // Mark the button as pressed delay(200);
      // Short delay to allow time for the reading to be taken
    } else { // If the button was already pressed
      PosIniServo = 0; // // Return to initial position
      (0 degree) SM.write(PosIniServo);
      ButtonPressed = false; // Mark the button as released
      again delay(200);
    }
  }
}
```

10 APPENDIX B - SCHEMATICAL SCHEMATICAL OF CIRCUIT FOR AUTOMATING THE CATAPULT

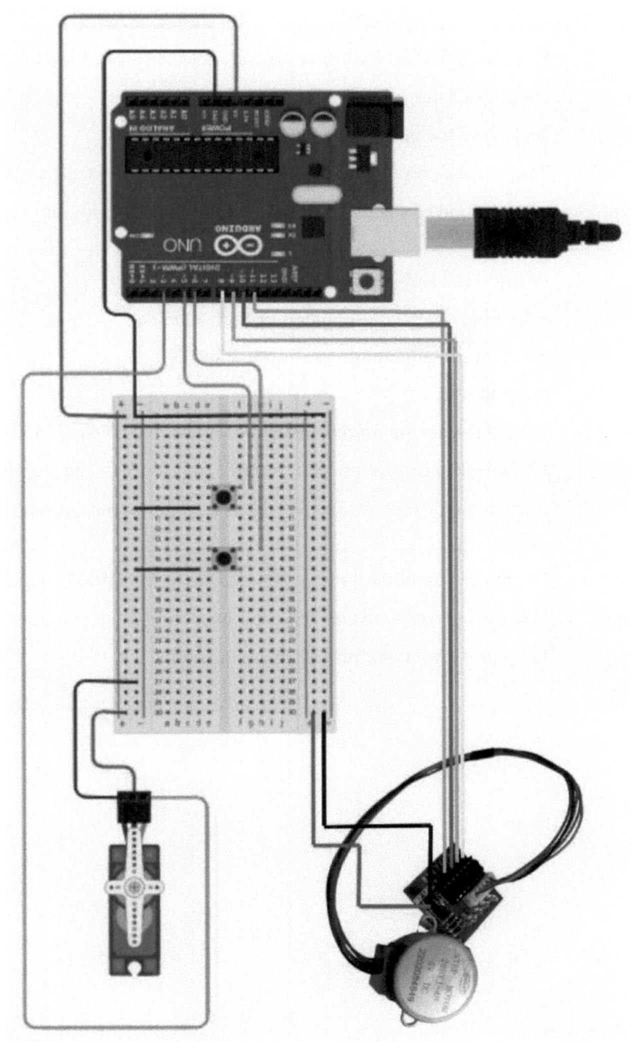

Source: The author

11 APPENDIX C - PROPOSED FLOWCHART FOR BUILDING THE CATAPULT

Meetings	Task
1st Meeting 2 lessons of 50 min	Cut the sticks and barbecue skewers into the sizes shown; sand the cut ends of the sticks; Assemble the base of the catapult; These tasks correspond to steps 1, 2 and 3 of the step-by-step construction of the catapult without automation/programming
2nd Meeting 2 lessons of 50 min	Position the rod on the catapult shaft; Assemble the catapult launch limiter; Finalize the catapult launch limiter; Finalize the catapult rod; Thread the elastic onto the rod and fit it into the release limiter; These tasks correspond to steps 4, 5, 6, 7 and 8 of the step-by-step construction of the catapult without automation/programming.
3rd Meeting 2 lessons of 50 min	Cut out the cardboard rectangles and mark the notable angles on them; This task corresponds to step 9 of the step-by-step construction of the catapult without automation/programming.

Printed by Books on Demand GmbH, Norderstedt / Germany